STUDENT UNIT GUIDE

A2 Geography
UNIT 2684

Specification **A**

Module 2684: People and Environment Options

Michael Raw

Philip Allan Updates
Market Place
Deddington
Oxfordshire
OX15 0SE

tel: 01869 338652
fax: 01869 337590
e-mail: sales@philipallan.co.uk
www.philipallan.co.uk

© Philip Allan Updates 2003

ISBN-13: 978-0-86003-748-4
ISBN-10: 0-86003-748-7

All rights reserved; no part of this publication may be reproduced, stored in a retrieval system, or transmitted, in any form or by any means, electronic, mechanical, photocopying, recording or otherwise without either the prior written permission of Philip Allan Updates or a licence permitting restricted copying in the United Kingdom issued by the Copyright Licensing Agency Ltd, 90 Tottenham Court Road, London W1T 4LP.

Five questions in the Question and Answer section are reproduced from past unit tests by permission of OCR. They are: Question 3 (January 2003), Question 5 (January 2002), Question 6 (June 2002), Question 7 (June 2002) and Question 8 (June 2002).

This guide has been written specifically to support students preparing for the OCR Specification A A2 Geography Unit 2684 examination. The content has been neither approved nor endorsed by OCR and remains the sole responsibility of the author.

Printed by MPG Books, Bodmin

Environmental information
The paper on which this title is printed is sourced from managed, sustainable forests.

OCR (A) Unit 2684

Contents

Introduction
About this guide 4
The aim of Unit 2684 4
Assessment pattern 6

■ ■   ■

Content Guidance
About this section 14
Geographical aspects of the European Union (EU)
Key questions 15
Key questions answered 15
Managing urban environments
Key questions 22
Key questions answered 23
Managing rural environments
Key questions 29
Key questions answered 30
Hazardous environments
Key questions 35
Key questions answered 36

■ ■ ■

Questions and Answers
About this section 42
Q1 Geographical aspects of the European Union (EU) (I) 43
Q2 Geographical aspects of the European Union (EU) (II) 46
Q3 Managing urban environments (I) 49
Q4 Managing urban environments (II) 52
Q5 Managing rural environments (I) 55
Q6 Managing rural environments (II) 59
Q7 Hazardous environments (I) 62
Q8 Hazardous environments (II) 65

Introduction

About this guide

This guide has been written to help you prepare for OCR Specification A A2 Geography **Unit 2684: People and Environment Options**. People and Environment Options is the synoptic unit that you sit at the end of your A-level course. It aims to assess your knowledge and understanding of AS and A2 geography as a whole, as well as the specific content of Module 2684. This means that examination questions are necessarily broad, providing you with opportunities to show understanding of the different aspects of geography you have studied. The book is in three sections:

- an **Introduction**, which explains the concept of synopticity, discusses the style of essay questions, and summarises the command terms used most frequently. This section concludes with advice on planning synoptic essays, a list of some general themes explored by synoptic questions, and the mark scheme used by examiners to assess your work. The last item is particularly important because it details exactly what the examiners are looking for.
- a **Content Guidance** section, which is more specific, and deals with each of the four option topics offered in this module. You should focus on the two option topics you have studied. For each of your options, familiarise yourself with the key questions and follow through the synoptic links to work you have done at AS and A2. The synoptic connections are emboldened, and indicate the areas of the specification outside this module that you should revise.
- a **Questions and Answers** section, which includes eight specimen questions (two for each option topic). Each of these is accompanied by an examiner's analysis of the question, with advice on how to tackle it. Read this section carefully and think about the synoptic opportunities: how could you develop them to show understanding of the connections between, for example, the physical and human environments? Each question concludes with a student's answer, complete with examiner's comments and marks.

The aim of this guide is to improve your preparation for the People and Environment module. In a real sense, the more you know about the module — its content, its key themes, its structure and its methods of assessment — the better you are likely to perform in the final examination.

The aim of Unit 2684

Unit 2684 is very different from the other units of the OCR(A) Specification. This is because it requires you to 'show understanding of the connections between the different aspects of geography represented in the specification'. Examiners refer to this as **synoptic assessment**. To understand this concept, it might help if you think of the synoptic charts you studied for AS atmospheric systems. A synoptic chart brings together lots of information about the current weather — precipitation, pressure patterns, wind direction and so on. In the People and Environment Options unit, the essays you write are similar to synoptic charts: they should bring together your knowledge and understanding of different parts of the geography course in order to answer specific questions.

This synoptic requirement can be approached in a number of ways. Your specification uses a simple approach: discursive essays. In this module, you study two options chosen from: Geographical Aspects of the European Union (EU); Managing Urban Environments; Managing Rural Environments; and Hazardous Environments. The essay questions in each option are broad and allow you to show your understanding of the interconnected nature of geography. Answers that are truly synoptic draw on both (a) the content of the options and (b) knowledge and understanding from your study of geography at AS and A2. Geography as a subject lends itself to this synoptic approach. This is because its focus on themes such as people–environment relationships and spatial relationships requires an understanding of physical systems (atmosphere, ecology, hydrology, lithosphere) and human systems (demography, economy, society).

The table below lists some of the synoptic connections between the four options of Module 2684 and the rest of the AS/A2 specification.

Option	Synoptic connections
Geographical Aspects of the European Union (EU)	• Systematic knowledge and understanding of geographical (e.g. environmental, human and physical) principles, processes and decision-making within the EU • Environmental and economic impacts of transnational issues on the EU
Managing Urban Environments	• Knowledge and understanding of urban settlement and population patterns, processes and problems • Knowledge and understanding of aspects of the urban physical environment (e.g. hydrology, industry, land use, pollution and services)
Managing Rural Environments	• Knowledge and understanding of rural settlement and population patterns, processes and problems • Knowledge and understanding of aspects of the rural physical environment (e.g. agriculture, climatology, hydrology and tourism)
Hazardous Environments	• Knowledge and understanding of population patterns, hazardous environments (e.g. coastal, fluvial and glacial), agriculture and tourism • Knowledge and understanding of tectonics, and atmospheric and hydrological processes relating to hazards

The diagram below illustrates the kind of connections you could make when answering a specific question in this unit. (This example is on Hazardous Environments.)

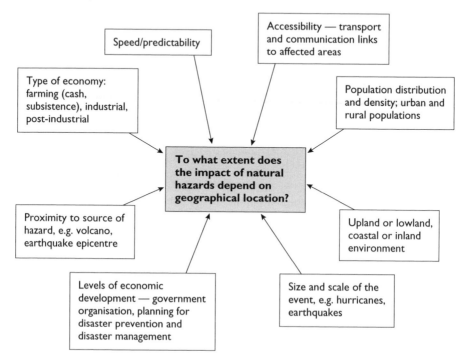

Another synoptic example is the depletion of commercial fisheries in the EU. To understand this issue, you first need to know how the Common Fisheries Policy operates. Then you need to understand the ecology of continental shelf areas, food chains, use of sustainable resources, the economics of commercial fishing and the economies of regions dependent on fishing. Similarly, if we consider the impact of hurricanes on communities in the Caribbean or Bangladesh, it is clear that we must understand the interaction between physical systems, such as the atmosphere and the oceans, and human systems, including economies, levels of development, population distribution and population density.

Assessment pattern

The People and Environment Options unit comprises one $1\frac{1}{2}$ hour paper, which offers three open-ended essay questions on each of the four synoptic options. You must answer two questions chosen from different options. The unit is worth 120 uniform marks, or 20% of the total marks for A-level, and each essay is marked out of 60.

OCR (A) Unit 2684

Style of essay questions

The essay questions in this unit have a number of features:
- They are discursive. This means that you will have to present, consider and analyse conflicting arguments and ideas.
- They require evaluation of a statement, argument or viewpoint. This means that any discussion should include a value judgement. Remember that there are no right or wrong answers. In this sense, the questions are open-ended. However, you must be sure that your judgement is consistent with your discussion.
- They are broad-based, inviting answers that draw on and apply (a) knowledge and understanding of the People and Environment Options and (b) knowledge and understanding of connections with other aspects of geography (i.e. physical, human and environmental) you have studied in your AS/A2 course.

Although question themes are derived from the synoptic options, effective answers must range beyond the content of each option. This is the essential requirement of synopticity and adds an extra demand to the questions in this unit. A typical question from Unit 2684 is examined below.

> **'Current rates of urban growth and urbanisation in LEDCs are unsustainable.'** Illustrate and **discuss the accuracy** of this statement.

- Knowledge and understanding of human geography (urban and rural population growth, and rural–urban migration) at AS
- Impact of urban growth/urbanisation on environmental systems (e.g. the atmosphere, water resources and land)
- Knowledge and understanding of the atmosphere and hydrology at AS
- The concept of sustainability in urban contexts (air pollution, pollution of water supplies, water resource and landfill shortages etc.) from People and Environment Options
- Discussion and evaluation

Synoptic themes

Essay titles in the People and Environment Options unit are necessarily broad in order to allow you to make the crucial connections with other aspects of geography in the specification. The following table lists some examples of themes that allow this synoptic approach.

Theme	Examples
The impact of human activities on the physical environment	Loss of habitat and environmental pollution caused by intensive agriculture; overfishing and the depletion of fish stocks in the EU; air pollution in urban environments
The influence of the physical environment on human activities	Tidewater locations for heavy industry; tourism and recreation in coastal environments and national parks; farming systems in the sub-arctic and tropical rainforest
The changing relationship between people and the physical environment	The decline of agriculture and changing relationships between people and the land in rural regions; growing use of the physical environment for recreation and tourism; an awareness of the need to work with, rather than against, environmental systems (e.g. coastal management)
The relationship between government and environment	National and supranational governmental initiatives (e.g. greenfield and brownfield sites), combating atmospheric pollution (e.g. clean air acts), conservation of habitats and landscapes
The interaction of people and the physical environment — its impact on different scales	Atmospheric pollution occurs on a local scale (e.g. vehicle exhaust in urban areas), national and international scales (e.g. acid rain) and on a global scale (e.g. global warming)
The spatial expression of geographical systems on different scales	Spatial variations in prosperity and human well-being on urban, regional and international scales
The sustainable use of environmental resources	Sustainable use of: biotic resources such as fish and timber; soils; resources for recreation and tourism; resources which support large urban areas (e.g. water, land and the atmosphere)
The importance of geographical location and accessibility for human activities	Access to markets for producer services; accessibility and core/periphery contrasts in wealth; access to employment and the location of squatter settlements in cities in LEDCs

Responding to command terms

Command words and phrases in examination questions tell you exactly what you have to do. You must respond precisely to their instructions: 'Discuss', for example, is a very different requirement from 'Describe' or 'Explain'. Ignoring the instructions of command words and phrases is a basic error of examination technique and is, perhaps, the most common cause of underachievement. Take, for example, the following question:

To what extent can recreation and tourism solve the economic problems of rural areas in MEDCs?

This question asks you to assess the capacity of one economic activity (in this case recreation and tourism) to solve rural economic problems. It invites discussion. But effective discussion is only possible by considering the role of economic activities such as agriculture, industry and services, as well as recreation and tourism. A description and explanation of recreation and tourism as a vehicle for economic growth would not provide a relevant answer. If you adopted this approach you would essentially be answering a different question, namely:

OCR (A) Unit 2684

How could recreation and tourism help to solve the economic problems of rural areas in MEDCs?

A broad approach, looking at several economic activities but containing neither discussion nor evaluation, would be equally inappropriate.

The requirements of the standard commands are listed below.

Command term/phrase	Requirements
Examine the view that...	Analyse and evaluate the validity, accuracy or appropriateness of a statement or a contention
To what extent...?	Analyse and evaluate the validity, accuracy or appropriateness of a statement or a contention
Assess...	Analyse and evaluate the importance or value of, for example, a process or factor
Discuss...	Consider the connected arguments surrounding an issue or debate and make reasoned judgements
How important is...?	Evaluate the importance of a factor, process or argument compared with others
How far do you agree with...?	Discuss and evaluate a statement
How successful is...?	Discuss and evaluate the success of a process or policy
How useful is...?	Discuss and evaluate the usefulness of a process, policy or theory
How and with what success...?	Describe the nature of a policy or initiative and evaluate its success

Think synoptically: plan your answers

A successful essay begins at the planning stage. All the questions on the paper are unseen and you will need to think quickly and make decisions about synoptic links and the content and scope of your answer. You should set aside 5 minutes for this crucial planning stage, and give yourself 40 minutes' writing time.

Bear in mind that all questions in this unit make the same demands: they are discursive, evaluative and invite you to show knowledge and understanding of other aspects of geography covered by the specification. Start by thinking synoptically! Before you consider the content and structure of your answer, draw a simple spider diagram to remind yourself how the question connects with other aspects of geography you have studied. This will focus your attention on the synoptic requirement of this unit. You might also include the diagram as part of your answer: it will at least signal to the examiner that you are thinking synoptically.

Once you have identified the appropriate synoptic connections, move on to your formal essay plan. This should comprise three sections:

(1) Introduction. This should define important terms used in the question and give some indication of the structure of the discussion to follow. It should be brief: three or four sentences are normally enough.

(2) Substance and discussion. List the key points for discussion and select, where appropriate, examples of specific places and environments. Try to ensure that each discussion item includes some evaluation. Consider where you might use sketch maps, charts and diagrams as an alternative to text.

(3) Conclusion. This should be a brief summary or synthesis of the discussion and a clear statement and justification of your position on the issue raised by the question. Ideally, you should know what the conclusion of your essay will be before you start writing your answer.

An example of an essay plan is provided below.

> **'Urban population growth is inevitable: urban sprawl is not.' Discuss the validity of this statement.**
>
> **1 Introduction**
> - Definition of urban sprawl: the expansion of an urban area into the adjacent rural and semi-rural areas.
> - Introduce the dichotomy between MEDCs, where urban populations have declined in the past 40–50 years and where strict planning control limits urban sprawl, and LEDCs, where urban populations are growing rapidly and where there are few constraints on urban sprawl.
>
> **2 Substance and discussion**
> - Urban population growth in LEDCs and MEDCs: causes, rates.
> - Critical analysis: definition of urban populations; variations/anomalies within MEDCs and LEDCs. Examples: contrasts between cities in the USA — northeast and south/south-west.
> - Urban sprawl and its environmental impact: loss of amenities, farmland, wildlife habitats etc. Contrasting attitudes to urban sprawl. Urban physical expansion in LEDCs (e.g. Mexico City) and in some MEDCs (e.g. Los Angeles, Las Vegas).
>
> **3 Conclusion**
> - Synthesis: urban population growth is not inevitable. The experience of LEDCs is very different from MEDCs. Urban population growth has largely ended in most MEDCs, though there are exceptions, especially outside Europe. Urban sprawl can be controlled by planning, though its effectiveness depends on values, attitudes and resources. Even in MEDCs in Europe, some degree of urban sprawl is accepted.
> - Evaluation: final judgement on the statement in the question.

Marking synoptic essays

Examiners use a mark scheme based on four assessment objectives. For each of these, the examiner identifies the level of attainment of your answer, and then allocates a mark within that level. You should familiarise yourself with the criteria below: they give you the best guide to what examiners are looking for. Study the eight sample essays in the Question and Answer section to see how this mark scheme is applied.

OCR (A) Unit 2684

1 Knowledge of content		Marks
Level 4	Candidates show detailed knowledge in a synoptic context of appropriate themes, processes, specific environments and places drawn from different aspects of geography represented in the specification. They have an accurate knowledge of relevant concepts, principles and theories, and of a wide range of geographical terms.	7–8
Level 3	Candidates show clear knowledge in a synoptic context of appropriate themes, processes, specific environments and places drawn from different aspects of geography represented in the specification. They have a clear knowledge of relevant concepts, principles and theories, and of a range of geographical terms drawn from different aspects of geography represented in the specification.	5–6
Level 2	Candidates show basic knowledge of appropriate themes, processes, specific environments and places drawn from different aspects of geography represented in the specification, but the synoptic context may be implicit rather than explicit. They have a basic knowledge of relevant concepts, principles and theories, and of a range of geographical terms drawn from different aspects of geography represented in the specification.	3–4
Level 1	Candidates show some knowledge of appropriate themes, processes, specific environments and places drawn from different aspects of geography represented in the specification. They have some knowledge of some relevant concepts, principles, theories and geographical terms drawn from different aspects of geography represented in the specification. There is little synoptic overview.	0–2

2 Critical understanding of content		Marks
Level 4	Candidates show a detailed understanding of the synoptic content and connections between the different aspects of geography represented in the specification.	18–22
Level 3	Candidates show a clear understanding of the synoptic content and connections between the different aspects of geography represented in the specification.	12–17
Level 2	Candidates show basic understanding of the synoptic content and connections between the different aspects of geography represented in the specification.	6–11
Level 1	Candidates show only the most superficial understanding of the synoptic content and connections between the different aspects of geography represented in the specification.	0–5

3 Application of knowledge and critical understanding in unfamiliar contexts		Marks
Level 4	Candidates show understanding by applying their knowledge of the specification content and synoptic links, relevantly and where appropriate, at a range of scales. They evaluate arguments, ideas, concepts and theories clearly.	18–22
Level 3	Candidates show understanding by applying most of their knowledge of the specification content and make relevant synoptic links, where appropriate, at different scales. They provide sound evaluation of arguments, ideas, concepts and theories.	12–17
Level 2	Candidates show understanding by applying some of their knowledge of the specification content relevantly and attempt a basic evaluation. The synoptic context may be implicit rather than explicit.	6–11
Level 1	Candidates explain contexts using basic ideas and concepts, with little synoptic overview.	0–5

4 Communication		Marks
Level 4	Candidates use an appropriate range of communication skills fluently and in different formats; present information within a logical and coherent structure; where appropriate, synthesise information from a variety of sources; use spelling, punctuation and grammar with a high level of accuracy; and employ geographical terminology with confidence.	7–8
Level 3	Candidates use an appropriate range of communication skills clearly in different formats; present information within an effective structure; where appropriate, show some ability to synthesise information; use spelling, punctuation and grammar with reasonable accuracy; and use a range of geographical terms.	5–6
Level 2	Candidates use a limited range of methods to communicate knowledge and understanding; make some effort to structure their work; use spelling, punctuation and grammar with some accuracy; and have a basic knowledge of geographical terminology.	3–4
Level 1	Candidates use a limited range of methods to communicate knowledge and understanding; make only a basic attempt to structure their work; use spelling, punctuation and grammar with variable accuracy; and have only sparse knowledge of geographical terminology.	0–2

Content Guidance

This section provides a summary of key questions and synoptic connections for each of the four People and Environment options. This format encourages you to organise your learning and revision around topics that figure prominently in the final assessment (Unit 2684). Themes such as the relationship of people to the physical environment and the influence of distance on economic activities provide a structure around which to use your synoptic knowledge and understanding of the A-level specification as a whole.

Each option in this section is divided into two parts. First, there is a summary table, listing the key questions and pointing out possible synoptic connections with other parts of the specification. The second part develops and amplifies the content of the summary tables: it provides answers to the key questions, suggests possible examples and signposts opportunities for synoptic development.

You should familiarise yourself with the key questions and read through the 'key questions answered' sections. Then follow up the synoptic references emboldened in the text. Remember that a synoptic approach is essential for success in this unit. In order to achieve this, you must refer back to relevant class notes, hand-outs, articles, textbooks and websites that you have used at AS and A2.

OCR (A) Unit 2684

Geographical aspects of the European Union (EU)

Key questions

Key question	Synoptic connections
How valid are the concepts of core and periphery?	Relative location and accessibility; settlement, development and environmental resources (2681). Physical geography distorting the economic pattern. EU example (2684).
What influence do physical geography and natural resources have on the distribution of wealth at regional and continental scales?	Atmosphere, soils, hydrology, relief (2680). Urban growth (2681). The concept of natural resources; agriculture, industry and tourism (2683). EU example (2684).
What influence do physical geography and natural resources have on the economic status of specific regions?	Relief, climate, soils, hydrology, population change, urban growth (2680, 2681). Options that focus on economic geography, natural resources and unequal development in space (2683). EU example (2684).
What are the causes of regional economic problems in specific regions and how successful have governments been in tackling them?	Physical geography (2680); industrial decline (2683); rural decline (2681); rural population change (2681, 2684); remoteness, accessibility, congestion (2681). Causes of regional economic problems considered in several 2683 options. EU example (2684).
Outline the major transnational environmental problems in the EU. How successfully have governments tackled these problems?	Knowledge and understanding of physical systems of atmosphere, ecology, hydrology (2681). Pollution of atmosphere, land and water (2683). EU example (2684).
How important is government in influencing the geography of economic activity?	The role of government is considered in all 2683 options. EU example (2684).
To what extent do the benefits of foreign direct investment (FDI) and globalisation outweigh the costs?	Globalisation of industry, farming and tourism in 2683 options. EU example (2684).
To what extent are physical and biological resources managed sustainably in MEDCs?	Ecosystems and ecological resources (2680). Sustainability in urban areas (2681). EU fishing and agriculture, rivers and seas (2684).

Key questions answered

(Possible synoptic connections are shown in **bold**.)

How valid are the concepts of core and periphery?

Core and periphery are defined both economically and geographically. The development of core and periphery suggests an over-concentration of economic activity and wealth in a region or group of contiguous regions and an under-concentration elsewhere. There is also an argument that prosperity in the core is at the expense of the periphery (**backwash effects**). The core is usually highly urbanised, the periphery more rural. Core and periphery often have a long history based on the relationship between the **physical environment**, **natural resources**, **relative location** and economy. In the EU, measures such as regional GDP per capita, levels of employment, regional economic growth, population growth and **net in-migration** suggest that the EU has an economic and geographical core area, bounded by cities such as London, Frankfurt, Milan and Paris. The periphery comprises eastern Germany, northern Britain, western Ireland and the Mediterranean basin.

Myrdal's model of **cumulative causation** explains the growth of the core. A series of **virtuous growth circles** favour investment, employment and the accumulation of wealth at the core. This growth leads to further growth. The fastest growing industries (i.e. those based on services and high technology), requiring highly skilled workers and good global communications, have become concentrated at the EU's core. The core becomes increasingly **accessible**; the periphery is, by comparison, **geographically isolated**.

But the **economic** and **geographical core** and the periphery are not necessarily interchangeable. Less prosperous regions may be situated centrally within the geographical core, while belonging to the economic periphery. EU examples include declining industrial regions such as Nord-Pas-de-Calais, southern Belgium, the Ruhr and South Yorkshire. Regions of relatively rapid growth and increasing prosperity may be found in the geographical periphery. These regions may have **comparative advantages** such as natural resources and skilled workforces for economic activities like tourism and high-tech industry. Examples of such regions in the EU include Finland, Ireland and Andalucia.

Core and periphery are simple concepts for understanding the broad geographical distribution of wealth and prosperity. However, they are highly generalised and over-simplified, as the example of the EU shows.

What influence do physical geography and natural resources have on the distribution of wealth at regional and continental scales?

Relative location, **accessibility** and **external economies**, as well as physical geography and natural resources, influence the geographical distribution of wealth at regional and continental scales. Upland regions, with steep **slopes**, thin **soils** and cold, wet **climates**, provide only a limited resource base for agriculture. Some of the least prosperous regions in the EU, such as northern Scotland, France's Massif Central and southern Italy, are dominated by uplands. **Inaccessibility** and **remoteness** add to the problems of upland regions. **Drought** is a feature of climates in southern Europe. Without irrigation, the hot dry summers of the Mediterranean basin limit agricultural productivity. The result is some of the poorest regions in the EU.

In contrast, many prosperous lowland regions have a rich resource base for farming (e.g. the Paris basin and East Anglia) and are highly accessible.

What influence do physical geography and natural resources have on the economic status of specific regions?

Many factors influence the prosperity and economic status of regions: **accessibility**, **economic/employment structure**, human resources (e.g. skills, quality of workforce), **acquired advantages**, **external economies**, **physical geography** and natural resources.

Prosperous regions such as London, Randstad and Paris in the EU have a number of common characteristics:
- the size and quality of their workforces
- their situations as **international communication hubs**
- their ability to attract **foreign direct investment**
- their economic structure, heavily biased towards prosperous **producer services** and **high-technology industry** and research and development
- their well-developed **infrastructure** of higher education institutions, ports and motorways

Physical geography and natural resources may have given such regions an **initial advantage**, but today their prosperity is based on man-made advantages.

Some regions owe their prosperity to physical geography and natural resources. Regional economies based on **recreation and tourism** often have strong links with climate, relief, landscapes and beaches. Prosperous sub-regions such as the Costa del Sol, the Cote d'Azur and western Austria benefit from physical resources which promote recreation and tourism. The prosperity of rural regions such as the Paris basin and East Anglia depends partly on physical geography, and resources such as **climate** and **soils** which favour **intensive agriculture**. Norrbotten's relative prosperity is based on resources such as iron ore, hydroelectric power (HEP) and forestry.

Less prosperous regions may suffer from **isolation** and **inaccessibility** (such as northern Finland, the Western Isles, the Greek islands). Some regions may have structural problems — for example, manufacturing industries that are in decline. Examples include northeast England, the Saarland, Silesia and Nord-Pas-de-Calais. In some regions, decline in the natural resource base (e.g. coal in South Yorkshire, fish stocks off northeast Scotland) may lead to unemployment, **depopulation** and low rates of economic growth. Rural regions with a poor resource base, such as southern Spain and the northern Pennines, will also experience depopulation and economic decline.

Physical geography and natural resources have an important influence on regional prosperity. However, other factors such as accessibility, economic structure, human resources and historical growth also exert a strong influence.

What are the causes of regional economic problems in specific regions and how successful have governments been in tackling them?

The causes of economic problems in specific regions are covered in the second and third answers in this section.

Government regional policy operates at national and international levels in the EU. Current regional policies operated by the UK government centre on financial incentives (regional selective assistance) designed to attract investment and create and/or safeguard jobs in assisted areas. Small businesses are eligible for regional enterprise grants. Government regional policies date back to the 1930s. The least prosperous regions in the 1930s largely remain the least prosperous today. That is not to say that regional policies have failed — without regional policies, the inequalities could be even greater today.

The EU operates its own regional policies on an international scale. Money for regional assistance comes mainly from the European Regional Development Fund. The EU defines two types of assisted area: Objective 1 and Objective 2. The least prosperous regions qualify for Objective 1 assistance (e.g. Cornwall, south Wales, Merseyside, Andalusia and eastern Germany). Objective 2 areas have been affected by industrial decline, **over-dependence on industry** and have high rates of unemployment. Investment in these regions aims to create jobs, improve **infrastructure**, education, skills, and the quality of the **physical environment**. In Objective 1 regions, the scale of investment (matched by government funding) is huge. There is evidence that in some regions (e.g. Ireland and Andalusia) this regional assistance has had a significantly positive impact.

Outline the major transnational environmental problems in the EU. How successfully have governments tackled these problems?

Transnational environmental problems extend across political frontiers. In the EU, major transnational environmental problems are pollution of seas, pollution of international rivers and acid rain. Solving these issues requires international cooperation between the governments of member states.

Transnational problems are often the most difficult to solve. The atmosphere, international rivers and seas have **common ownership** and it may not be in the interests of individual states to control pollution while others continue to pollute.

Pollution of the North Sea occurs through discharges from rivers of chemical and organic waste from **industry**, **agriculture** and **urban areas**. There is also atmospheric fall-out of pollutants, and oil pollution from offshore rigs and tankers. Pollutants (e.g. heavy metals) enter **food webs** through plankton and accumulate in heterotrophs at each **trophic level** in marine **ecosystems**. **Eutrophication** is a particular problem in areas where untreated sewage and **nitrate fertilisers** are found in high concentrations. Solid pollutants may also affect the attractiveness of coastlines for **recreation** and **tourism**.

Similar pollution problems affect international rivers such as the Rhine — in particular, fish populations and abstractions for water supply.

Acid rain (including dry deposition) destroys forests, **acidifies soils** and makes lake and river water toxic for aquatic life. The main cause of acid rain is burning **fossil fuels** in power stations and emissions of sulphur dioxide. Prevailing **southwesterly winds** transport pollutants from industrialised areas of the UK, Germany and France to Scandinavia.

Governments have had some success in tackling all three transnational environmental problems. International agreement has been helped by the EU and by the polluting countries having similar levels of development. North Sea conferences in the 1990s made real progress and today the North Sea is managed **holistically** (i.e. as an ecosystem). The aim is to secure the sea's health and **biodiversity** and manage its fish stocks **sustainably**.

Huge improvements have been made in cleaning up the River Rhine. The Rhine Action Programme (1987), agreed by the countries through which the Rhine flows, covers all aspects of the Rhine ecosystem, and has been instrumental in reducing pollutants from **industry** and **agriculture**, improving water quality and biodiversity.

The acid rain problem has also been tackled positively through international cooperation. A whole raft of initiatives has been launched since the mid-1980s aimed at reducing sulphur in fuel oils, and emissions of sulphur dioxide and nitrogen oxide. Although there is still much to do (around 7000 Swedish lakes are only kept alive by liming), significant reductions have occurred in emissions of acidifying substances.

How important is government in influencing the geography of economic activity?

The geography of economic activity covers **agriculture**, **manufacturing industry** and **services** (including **tourism**). Government plays an important role in the location and functioning of these activities within the EU.

The EU has a common policy for agriculture. The Common Agricultural Policy (CAP) is arguably the most important influence on agriculture in the EU. Using subsidies for production, area and livestock, and grants for farm improvements and less favoured areas, the CAP has a profound influence on agricultural land use, farming methods, **conservation of ecosystems**, **habitats**, historic **landscapes** and **wildlife**. In addition to the CAP, individual governments have their own schemes (e.g. Country Stewardship in the UK). However, agriculture is also affected by many non-governmental influences, especially the **physical environment** (**climate**, **relief**, **soils**) and **access to markets**.

Governments influence the location of manufacturing industries through policies (at both international and national scales) designed to assist less prosperous regions. Grants are available to create new industries and maintain existing industrial employment in assisted areas. The EU defines Objective 1 and Objective 2 areas: many regions eligible for assistance under Objective 1 and Objective 2 have suffered serious **deindustrialisation** in the past 20 or 30 years. In the UK, manufacturers in assisted areas receive regional selective assistance. Individual governments may also secure **foreign direct investment** (e.g. from the USA, Japan and South Korea) by offering financial incentives to foreign **transnational corporations**. But the geography of manufacturing is strongly influenced by other non-governmental factors such as **economies of scale**, **labour**, **markets**, **transport** and **industrial inertia**.

Most government incentives available to manufacturing industry through regional policies are also accessible to **service activities** such as finance, banking, accounting, advertising, legal services and call centres. The influence of government depends how **footloose** these activities are. Many **producer services** rely on external economies which are only available in the largest urban agglomerations. In contrast, services such as call centres are more footloose and are therefore likely to be more responsive to government policies.

Although tourism and recreation are resource-based, governments can still influence their geography. Grants are available to farmers with the potential to **diversify** into tourism. **National parks** are government-financed, and run-down resorts may be eligible through regional policies for government financial assistance.

At a local scale, the siting of new factories, science parks or business parks is entirely under the **development control** policies of **planning authorities**.

To what extent do the benefits of foreign direct investment (FDI) and globalisation outweigh the costs?

FDI is overseas investment usually by **transnational corporations** (TNCs). Both manufacturing and service industries generate FDI — for example, a Japanese car manufacturer establishing a **transplant** in France or an American hotel chain building new hotels in the UK. FDI generated by TNCs is the force behind the globalisation of economic activity, that is, the establishment of production functions and markets worldwide.

FDI and globalisation have advantages and disadvantages for both countries/regions and companies. The advantages for countries/regions include:
- creation of employment
- new investment attracting further investment from **related/linked activities**
- trade (more exports/fewer imports)
- **improvements in infrastructure** (e.g. roads) to service new factories

The disadvantages are:
- lack of control, with the company's HQ located overseas
- **disinvestment** — companies may close plants/offices and shift production/activities overseas where costs are lower (e.g. call centres from the UK to India)
- creation of **branch plant economies**, with routine production offering low-skilled jobs lacking research and development functions

For companies, the advantages of FDI and globalisation include:
- **lower production costs** (labour, sourcing materials, taxes and environmental controls)
- **economies of scale**, with access to larger markets
- avoiding **tariff barriers to trade**

The disadvantages are fewer, but might include political instability (especially in some LEDCs) and charges of exploiting cheap labour and lax environmental controls.

To what extent are physical and biological resources managed sustainably in MEDCs?

Sustainable management means using resources in a way that does not compromise the needs of future generations. In other words, sustainable management does not have negative environmental effects. Relevant physical resources include the atmosphere, seas, soils, water, landscapes and habitats. Biological resources include plants and animals.

Physical resources that support agriculture often have not been managed sustainably. Intensive agriculture in the EU has degraded soils and resulted in: **soil erosion** by wind and water; contaminated **groundwater** and **surface water supplies**; destruction of wildlife habitats such as **wetlands**; and alteration of historic **cultural landscapes**. **Overgrazing** in upland regions has also contributed to accelerated soil erosion. Industrial activities have polluted watercourses, shallow seas and the atmosphere, contributing to problems such as **global warming**, depletion of the **ozone layer** and **acid rain**. **Leachates** from landfill sites and old mine workings have also had polluting effects on watercourses.

Biological resources have also suffered from unsustainable use. Fish stocks in the North Sea and the Grand Banks have been **overfished** to the point where some species may not recover. **Biodiversity** has declined as a result of intensive farming, with significant reductions in many bird, insect and wildflower populations. Habitats that support diverse wildlife, such as ancient woodlands, downlands and traditional hay meadows, have also suffered, largely due to **intensive farming** and **urbanisation**.

Following the **Rio Summit** of 1992, governments and government agencies have given the sustainable management of resources higher priority. In the EU, the CAP has been reformed and more emphasis placed on reducing intensive farming, partly for environmental reasons. Schemes such as set-aside, Environmentally Sensitive Areas and area payments have helped fuel this change. The Common Fisheries Policy (CFP) also aims to manage EU fish stocks sustainably, though with limited success. Pollution of rivers, seas and the atmosphere are being tackled, in some cases with conspicuous success (e.g. acid rain in Scandinavia, the North Sea and major European rivers). There has also been important growth in the **alternative energy sector**. But despite this, the **Kyoto Protocol** has not received unanimous support from MEDCs. Carbon dioxide emissions will continue to rise globally, threatening unprecedented **climate change**, disruption to **ecosystems** and **rising sea levels**.

Managing urban environments

Key questions

Key question	Synoptic connections
What is the nature of urban housing problems in MEDCs and LEDCs?	Housing stock, urban land use and urban morphology (2681).
What are the connections in LEDCs between housing problems and economic, social and political systems, rural development, demography and environmental hazards?	Rural–urban migration in LEDCs (2681). Natural hazards (2684). Management of urban environments (2684).
How have the urban housing problems of MEDCs and LEDCs been tackled, and with what success?	Housing stock, urban land use and urban morphology (2681). Management of urban environments (2684).
How and why do poverty, social exclusion and multiple deprivation within and between urban areas in MEDCs have an uneven geographical distribution?	Spatial segregation of economic, social, ethnic groups in cities; urban population distribution and density; urban housing stock; physical, environmental, negative externalities; urban land use (2681). Local climates (2680, 2683). Manufacturing and negative externalities (2683). Management of urban environments (2684). EU and regional inequalities (2684).
What are the causes/effects of urban sprawl? How and with what success has urban sprawl been managed?	Sub-urbanisation and ex-urban growth; urban–rural migration and counter-urbanisation (2681). Manufacturing and environment (2683). Services: out-of-centre provision (2683). Impact of human activities on hydrological systems and ecosystems (2680). Managing urban environments (2684).
What are the causes/effects of urban growth and urbanisation, and how sustainable is this growth?	Contemporary urbanisation in LEDCs; contemporary urban growth in MEDCs (2681). Environmental impact of urban growth; atmosphere, hydrology, ecosystems (2680). Managing urban environments (2684).
What are the causes/effects of traffic congestion in urban areas? How successful have strategies been to tackle the causes and effects of congestion?	Contemporary urbanisation in LEDCs and urban growth in MEDCs (2681). Environmental and economic consequences of congestion; air pollution (2680, 2683). Sustainable cities (2684).
What are the causes/effects of pollution in urban areas, and how successful have strategies been to contain them?	Problems of urban growth in MEDCs (2681). Local energy budgets; human activities in drainage basins (2680). Applied climatology (2683). Manufacturing (2683). Managing urban environments (2684).

Key questions answered

(Possible synoptic connections are shown in **bold**.)

What is the nature of urban housing problems in MEDCs and LEDCs?

Urban housing problems include: **housing availability**, **housing quality** and **housing costs**. These problems are common to both MEDCs and LEDCs — the differences are in their extent and severity.

Housing shortages in many MEDCs, including the UK, will increase in future. Increases in the number of households, caused by **societal** and **demographic changes** (rising divorce rates, more single, young-person households, increased life expectancies) rather than population growth, will result in a soaring demand for housing in the next 20 years. This trend has an important policy implication for urban management: should new housing developments occupy **greenfield** or **brownfield sites**? By comparison with MEDCs, housing shortages in urban areas in LEDCs are more acute. For the past 50 years or more, urban authorities in LEDCs have been unable to provide **low-cost housing** to accommodate rapidly growing populations. Even where such housing schemes have been developed, their location and relatively high rentals have proved unattractive. The importance of **squatter settlements** in most cities in LEDCs largely reflects the shortage of conventional housing. By building their own homes, the urban poor have, hitherto, largely solved their housing problems themselves.

The quality of housing stock in most MEDCs is also adequate. However, there are geographical variations in housing quality within cities. **Inner-city housing** often comprises high-density terraces, tenements, apartments and high-rise flats of variable quality. Much of this housing, particularly in continental Europe and North America, is in the **private rented sector**. In the UK, poor-quality housing is also found on edge-of-town local authority estates built in the 1930s, 1950s and 1960s.

Housing for the masses in many LEDCs often means self-built shacks in squatter settlements without essential services. Basic housing of this type may be preferred to conventional housing: it is cheaper, often closer to employment in the city, and when legal title is given to squatters, it will gradually be improved by the inhabitants. Ultimately, what started out as unplanned slums may be **upgraded**, provided with key services by the urban authorities and eventually **incorporated** into the city as mature suburbs.

In urban areas in MEDCs, a wide variety of housing types is available. High-income groups occupy the best-quality housing in neighbourhoods that maximise **positive externalities** (e.g. access to countryside, access to good schools). Low-income groups occupy lower-quality housing in neighbourhoods that may suffer **negative externalities** (e.g. high levels of crime and pollution). The cost of housing often varies regionally: areas of high-cost housing (e.g. southeast England) may present special problems for **in-migrants** and lower-income groups. In LEDCs, widespread poverty often means that rents, even in low-cost housing schemes, are beyond the reach of a large proportion of urban dwellers.

A2 Geography

What are the connections in LEDCs between housing problems and economic, social and political systems, rural development, demography and environmental hazards?

Shortages of/poor-quality housing and lack of services in urban areas in LEDCs have economic, social, political and environmental causes. Rapid **urbanisation** has overwhelmed many urban governments, resulting in severe housing shortages.

The most fundamental causes of housing problems in LEDCs are economic. The urban poor are unable to compete in the **housing market**. They have few options other than to occupy rented slums or rely on **self-help** and squat in spontaneous settlements. The absence of (a) conventional housing for the poor and (b) the most basic services reflects the limited resources available to city authorities.

Huge influxes of migrants into LEDC cities are an indication not only of a **lack of economic development in rural areas**, but also limited opportunities for social advancement, for example few educational opportunities, a rigid social hierarchy with few prospects of advancement, and social inequality (especially towards women). To these **push factors** can be added political instability and civil war. In the past, the political views of urban authorities often meant that squatter settlements were periodically bulldozed or razed, adding to the severe housing shortage.

Housing problems are often made worse by the **age structure** of **rural–urban migrants**. Young adults dominate migration flows and these groups and their children put added pressure on inadequate housing and service provision.

Environmental hazards, such as floods, droughts, earthquakes and volcanic eruptions, are responsible for some rural–urban migration. Within LEDC cities, the poorest people are often forced to live in places that are vulnerable to environmental hazards such as squatter settlements in flood-prone areas, or on steep slopes at risk from mass movements.

How have the urban housing problems of MEDCs and LEDCs been tackled, and with what success?

Housing shortages, sub-standard housing and urban sprawl have elicited a variety of planning responses in MEDCs. In the UK, France and other western European countries, the responses include slum clearance, high-rise flats, new towns and expanded towns. More recently, **reurbanisation** and the development of **brownfield sites** have been encouraged. These planning responses have had varying success. New towns have generally been successful, high-rise flats a disaster. Reurbanisation has been successful in some larger cities with a strong service base (e.g. Manchester, Newcastle, Birmingham and Leeds), but the demand for new homes cannot be accommodated entirely on brownfield land.

In many LEDCs, housing problems have been solved effectively by **self-help**. Squatter settlements are now seen as a positive response to housing shortages. Governments have granted squatters legal title to land, which encourages investment in, and improvement of, housing. Given time, residents **upgrade** their homes, and local authorities can provide **essential infrastructure**, such as electricity, sewerage systems and piped water. Low-cost solutions to housing problems, such as sites-and-services schemes and core housing, often sponsored by NGOs and the World Bank, have also proved successful in some cities.

How and why do poverty, social exclusion and multiple deprivation within and between urban areas in MEDCs have an uneven geographical distribution?

Poverty, social exclusion and multiple deprivation within urban areas are most frequent in inner cities and in local authority estates (in the UK). The latter, built within the past 50 years, have a predominantly peripheral location. Groups suffering from poverty/social exclusion/multiple deprivation have **low incomes** and a disproportionate number belong to **ethnic minorities**. Their geographical distribution within cities is most strongly influenced by housing types. Unable to compete in the **housing market** (because of low and/or unstable incomes, poor credit ratings etc.), they occupy the least desirable housing, usually in the rented sector. These housing areas are unattractive for a combination of reasons — the age of the housing, neighbourhood dysfunction such as high levels of crime, isolation from services and a poor **physical environment** (including pollution). The concentration and **segregation** of poor/socially excluded ethnic minorities may also reflect a positive choice, such as sharing a common language and culture, defence against a hostile host society and proximity to workplaces.

Poverty, social exclusion and multiple deprivation have spatial expression at a regional scale too. **Deindustrialised regions** in northern England, south Wales, Scotland and Northern Ireland have suffered from a decline in traditional employment sectors, such as coal mining, steel and heavy engineering. Examples include the decline of coal mining in east Durham and the demise of steel making in South Yorkshire. Similar **structural problems** in sectors like tourism and fishing have created pockets of poverty in regions such as Cornwall and northeast Scotland.

What are the causes/effects of urban sprawl? How and with what success has urban sprawl been managed?

Urban sprawl occurs when urban growth cannot be contained within a city and spills into the surrounding countryside. Urban areas may expand because of population growth. Today this is most likely in cities in LEDCs, where both **rural–urban migration** and **natural increase** create urban growth. In MEDCs, urban growth is more likely to result from societal change and rising living standards. Increased **life expectancy** and high rates of divorce fuel the demand for more **single-person households**.

Counterurbanisation also leads to urban sprawl. Rising standards of living and concomitant demands for larger houses with more extensive building plots, and a desire for a more rural life style, have **urbanised the countryside** around many large cities in MEDCs. Lax planning controls (e.g. in North America), where previously land shortage has not been a problem, also encourage urban sprawl.

Urban sprawl is facilitated in MEDCs (and in highly developed LEDC regions such as southeast Brazil) by efficient public transport systems and the **personal mobility** conferred by private car ownership.

Urban sprawl has **environmental** and **economic costs**. Loss of countryside to urban development destroys woodland, hedgerows, ponds and other wildlife habitats and ecosystems. There is also a loss of agricultural land and amenity areas. Urban sprawl increases the lengths of **journeys to work** and, where public transport is poorly developed, may contribute to traffic congestion and **air pollution**.

Attempts to contain urban sprawl include: **green belts** and **green sectors**, where strict development controls apply; new and expanded towns to accommodate overspill population; **reurbanisation**; and recycling urban land (i.e. brownfield sites). Green belts have been the favoured approach in the UK since the 1930s. They interrupt urban growth but cannot stop development leapfrogging the green belt and consuming the countryside beyond. Such is the scale of the expected housing demand in the UK in the next 15 years that significant encroachment of urban development into green belts is inevitable. Green sectors/axes permit urban growth along designated urban corridors where conservation may be less important and amenity value low. Investment in transport systems in these corridors provides good access to the city, reducing both journey times for commuters and pollution problems.

Brownfield development is consistent with the notion of **sustainable cities**. However, brownfield sites are more expensive to develop than **greenfield sites** and housing schemes on these sites will necessarily be at relatively high density. Reurbanisation and the attraction of people back to city centres has been successful in many UK cities which are large enough to support a wide range of services such as clubs, restaurants, pubs and theatres.

What are the causes/effects of urban growth and urbanisation, and how sustainable is this growth?

Urban growth and **urbanisation** in LEDCs result from **rural–urban migration** and the **natural increase** of urban populations. **Push and pull factors** explain rural–urban migration. In rural areas, push factors include poverty, lack of employment opportunities, limited educational provision, debt and landlessness. Out-migrants anticipate a better quality of life in urban areas and respond to the lure of the city and its pull factors. A large proportion of migrants are young adults and this fact, together with better standards of living and medical care, produces high rates of natural increase in cities. Because urban population growth exceeds rural population growth, there is a relative **urban shift** of population. This is urbanisation.

Urbanisation is no longer taking place in MEDCs. However, urban growth continues as the number of households and the demand for larger homes increases. This demand is driven by changes in family structure and rising living standards.

Rapid urban growth and urbanisation in LEDCs have created serious housing problems, job shortages in the **formal economy**, inadequate service provision and

environmental pollution. Millions of urban dwellers rely on **self-help** for housing and employment. Urban authorities have limited resources and often struggle to provide even the most basic services. Urban growth is so rapid and haphazard that pollution of water supplies and the atmosphere are routine, and there are major problems with water supply and the disposal of waste materials. In most LEDC cities, current urban growth is **damaging the environment** and the **resources** that support urban populations. This situation is **unsustainable**.

Urban growth in MEDCs has led to urban sprawl, loss of countryside, traffic congestion, air pollution, water shortages and problems of waste disposal. Sustainable growth is high on the agenda of urban management and includes **recycling** of waste and water, congestion charges, limits on environmental pollution and the reuse of brownfield sites.

What are the causes/effects of traffic congestion in urban areas? How successful have strategies been to tackle the causes and effects of congestion?

Traffic congestion in urban areas is the result of the increase in vehicle use outstripping the increase in road space. The **layout** of most cities pre-dates the invention of the motor vehicle and is unsuited to modern traffic conditions. Congestion causes increased **pollution** and imposes **significant costs** on national economies.

The huge growth of **private car ownership** in the past 50 years in MEDCs is the principal cause of traffic congestion — a situation often made worse by inadequate **investment in public transport systems**. **Counterurbanisation**, which diffuses urban populations (and makes public transport provision difficult), also contributes to increasing car usage and traffic congestion. Other contributing factors include changes in **shopping behaviour** (e.g. superstores and planned regional shopping centres, which rely on car-borne shopping) and the use of cars for journeys to school.

Among the strategies to tackle traffic congestion are: investment in new roads (ring roads, by-passes), promotion of alternative forms of transport (cycle lanes), investment in public transport systems (trams, light rail, guided bus lanes), incentives to use public transport (park-and-ride), road charging and bus lanes. The decentralisation of some economic activities from city centres can help to alleviate congestion. New roads appear to generate more traffic and drivers are reluctant to abandon their cars for public transport. More coercive measures seem inevitable. Road charging is a radical new approach to the congestion problem in the UK.

What are the causes/effects of pollution in urban areas, and how successful have strategies been to contain them?

The causes of **pollution** in urban areas are motor vehicles, industrial waste and domestic waste. Atmosphere, hydrosphere and land are affected by pollution. Motor vehicle exhaust gases are routinely responsible for **photochemical smog** in Los Angeles and poor air quality for **temperature inversions** in Mexico City in winter. Vehicle exhaust gases may also build up in **anticyclonic conditions** to produce severe pollution events in MEDCs (e.g. Paris and London). Poor air quality is associated with high incidences of **respiratory disease** (e.g. asthma, bronchitis and lung cancer) and high rates of **morbidity**. Residential areas close to heavy industrial complexes (e.g. steel and chemicals on Teesside) often show above-average rates of morbidity and ill health.

In LEDCs, many cities have no sewage treatment systems, with the result that pollution of watercourses is common and a source of **disease**. In Mexico City, airborne faecal particles (blown by strong winds from a dry lake bed) are a threat to human health, and again reflect the absence of sewage treatment in large parts of the city. Untreated water waste may be used to irrigate areas around cities, **contaminating land** with heavy metals, such as cadmium and boron. **Air quality in squatter settlements** is often poor because of numerous workshop activities, such as metal foundries and smelting. The lack of **organised refuse disposal** and the location of poor housing areas adjacent to rubbish tips cause pollution and further threats to human health.

Environmental problems in LEDCs do not take the priority they assume in MEDCs. Limited resources are often diverted to more urgent problems, such as housing, job creation and service provision. MEDCs have legislation to proscribe **emissions of atmospheric pollutants** (e.g. the burning of coal in residential areas) and discharge of effluent to streams and rivers. Emissions from motor vehicles may be monitored; lead-free petrol is promoted by giving it a price advantage over leaded petrol; cars may be banned from central areas at certain times. Similar financial incentives may be given to car owners who choose liquefied petroleum gas rather than petrol and diesel fuels. Since 1996, solid waste dumped in landfill sites has incurred a landfill tax. This helps to encourage recycling.

Managing rural environments

Key questions

Key question	Synoptic connections
How important is it to maintain the survival of traditional communities in the countryside in MEDCs?	Concepts of threshold and hierarchy (2681). Rural population changes in MEDCs since 1960 (2681). Second homes and holiday homes (2684). Rural landscapes and traditional economic activities (2683).
What has been the impact of intensive farming on rural environments/ecosystems in the past 40 years?	Hydrology, soils, ecosystems (2681). Habitats (2684). Agro-ecosystems (2683). EU farm policies (2684). Rural management (2684).
How successful have recent attempts been to establish a more sustainable agriculture?	Hydrology, soils, ecosystems (2681). Habitats (2684). Agro-ecosystems (2683). EU farm policies (2684). Rural management (2684).
Examine the importance of the demographic factors that have influenced population change in rural areas in recent years.	Population changes since 1960; depopulation, net out-migration, net in-migration, counter-urbanisation, remote and pressured rural areas (2681). The economic, social and environmental causes of population change (2680, 2681). The influence of the physical environment, resources and accessibility (2680).
How successful have government initiatives been in halting the decline of rural communities? How have changing settlement hierarchies affected rural populations in the past 40 years?	Population changes since 1960; counter-urbanisation, remote and pressured rural areas (2681). Rural management policies (2684). CAP policies (2683). Rural settlement in MEDCs; range, threshold, hierarchy (2681). Experience of settlements close to and remote from large urban areas (2681). Out-of-centre retailing and changing service provision in rural areas (2683). Rural management (2684).
What are the causes of rural service decline? How successfully has rural service decline been tackled?	Rural settlement in MEDCs; range, threshold, hierarchy (2681). Experience of settlements close to and remote from large urban areas (2681). Out-of-centre retailing and changing service provision in rural areas (2683). Rural management (2684).
To what extent are human activities in national parks sustainable? How successful have governments been in achieving sustainable management?	Hydrology, soils, ecosystems (2681). Sustainable tourism (2683). Rural management (2684).

Key questions answered

(Possible synoptic connections are shown in **bold**.)

How important is it to maintain the survival of traditional communities in the countryside in MEDCs?

Population changes in the countryside include **net out-migration**, **ageing populations** and **depopulation** in remoter rural areas. Population turnover in pressured rural areas within commuting distance of major urban centres results in changes in the economic, social and demographic structure of rural communities and loss of countryside.

Traditional rural communities are centred around agriculture and supporting services. **Agriculture is multi-functional**: it provides direct employment in the countryside, supports a range of services, creates and maintains cultural landscapes, and helps conserve wildlife. **The decline of agriculture** and contraction of employment in this sector, together with few alternative job opportunities, undermine **service thresholds** (e.g. shops, transport and schools). This makes rural areas even less attractive and creates a negative spiral which results in further population loss.

Agriculture is responsible for **cultural landscapes** that have high amenity and conservation value (e.g. moorland, downland, ancient woodlands and hay meadows). These landscapes support a variety of **habitats, ecosystems and wildlife**. Any retreat of agriculture would threaten these **plagioclimax communities**. The value of the countryside for **tourism** might also suffer if historic cultural landscapes are replaced by **natural succession** and wilderness.

What has been the impact of intensive farming on rural environments/ecosystems in the past 40 years?

Intensive farming describes a system with high inputs and high outputs per unit area. Modern farming in both upland and lowland areas has become more intensive in the past 40 years. Government policies subsidising food production have been the driving force behind intensification. Intensive agriculture has had adverse environmental effects on **soils**, **hydrology**, **ecosystems** and **landscapes**.

Soil erosion caused by **over-cultivation**, the use of chemical fertiliser at the expense of organic fertiliser and overgrazing are related to intensification. Removal of hedgerows exacerbates **wind erosion**; deterioration in **soil structure** by use of chemical fertilisers and heavy machinery facilitates wind and water erosion; and **overgrazing** in uplands destroys the vegetation cover and exposes soils to erosion.

How successful have recent attempts been to establish a more sustainable agriculture?

Sustainable agriculture is the use of resources that meets the needs of the present without compromising the ability of future generations to meet their own needs. Modern, intensive agriculture in MEDCs has often degraded biological resources. Problems with non-sustainable agriculture include **land degradation, soil erosion, pollution of rivers, groundwater and coastal zones, loss of biodiversity**, and the **destruction of historic cultural landscapes**.

Recent attempts have been made to achieve a balance between production and the conservation of resources. **Over-cultivation** and **overgrazing** were features of agriculture in many MEDCs in the 1960s, 1970s and 1980s. Unsustainable agriculture was fuelled by generous government subsidies for crop and livestock production, export subsidies and import tariffs. In the EU, incentives are offered to move away from **intensive** agriculture by reducing production subsidies (area payments rather than payments for crop yields or based on stocking densities). Environmentally sensitive areas (ESAs) promote traditional cultivation practices and help conserve wildlife and landscapes. Country stewardship schemes, set-aside, nitrate-sensitive areas, hedgerow incentive schemes and organic farming also promote conservation.

The shift towards area payments and set-aside schemes has stopped overproduction and the worst **environmental effects** in the EU. Set-aside encourages the growth of weeds, which provide food for insects and birds and promote **biodiversity**. ESAs help to conserve traditional landscapes and wildlife in areas such as the Lake District and Pennine dales. Woodland incentive schemes have a similar effect.

Examine the importance of the demographic factors that have influenced population change in rural areas in recent years.

Population change results from either **natural change** or **migration**. Natural change is the numerical difference between births and deaths. When births exceed deaths, **natural increase** occurs; an excess of deaths over births results in **natural decrease**. **Net migration** is the numerical difference between in-migration and out-migration. An excess of in-migrants results in a net migration gain; an excess of out-migrants produces a net migration loss. **Depopulation** may be caused by natural decrease, net migration loss or a combination of the two.

Depopulation is widespread in remote rural areas in MEDCs. **Ageing populations** caused by **selective out-migration** of young adults means small numbers of adults in the reproductive age groups and low crude birth rates. Aged populations also increase death rates. Regional examples include the Highlands and Islands of Scotland, France's Massif Central, and the US mid-west. **Counterurbanisation** is caused by migration and has been the driving force in the **population revival** of some remote rural communities such as mid-Wales.

In pressured rural areas close to major urban centres, counterurbanisation has been responsible for rapid population growth in the past 20–30 years. In the UK, counties adjacent to large urban centres, such as Buckinghamshire, North Yorkshire, Warwickshire and Cheshire have experienced counterurbanisation and growth rates of 10–20% during this period.

How successful have government initiatives been in halting the decline of rural communities?

Government initiatives in western Europe operate on national and international scales. The CAP supports agriculture partly because it is vital for the survival of rural communities. Money is available from its Guidance Fund for **less-favoured areas**, agricultural improvements and **diversification** from farming to **off-land enterprises**.

Less-favoured areas are mainly uplands, where the **physical resources for farming (climate, soil, relief)** limit production. Farm improvements include **amalgamation of holdings** and **irrigation**. Diversification may involve tourism and the conversion of farm buildings to new uses. These policies contribute to the survival of rural communities in upland regions such as the Highlands of Scotland and the north Pennines, and the intensification and modernisation of agriculture in regions such as Andalucia and Murcia. EU structural funds are also available in rural regions whose development lags behind the rest of the EU. Large, predominantly rural regions, such as Cornwall and Sardinia, qualify for generous assistance to develop new employment, education/training and infrastructure.

At the national level, the UK government has introduced a raft of initiatives aimed at rural regeneration. Grants are available in Rural Priority Areas to assist individual enterprises and integrated rural development. The England Rural Development Programme supports agricultural diversification, and the Countryside Agency administers schemes and grants to support rural services, including transport. Planners have concentrated new development in selected centres (key settlements) which have the potential to maintain/achieve the **thresholds** to support retail services, healthcare, education and transport. New housing for rural communities is focused in key settlements where services are most viable. **Market towns** (higher status service centres, with employment potential, essential infrastructure and threshold populations to support services) are pivotal to **rural regeneration**. Growth is diverted here and away from villages.

The decline of rural communities continues in many remoter rural areas, with net out-migration depopulation and loss of services. However, without government policies at national and international levels, the scale of decline would be much greater. In some rural regions (e.g. Skye), recent **counterurbanisation** has reversed trends that prevailed for most of the twentieth century.

How have changing settlement hierarchies affected rural populations in the past 40 years?

Settlement hierarchies are the ordering of settlements according to their functional status or **centrality**. Rural settlement hierarchies comprise hamlets (first-order), villages (second-order) and market towns (third-order).

The functional status of settlements is determined by the **threshold** requirements of their services, and the **range** of services found in competing centres. Improvements in personal mobility and road infrastructure in the past 40 years have extended the range of services provided by larger market centres. Meanwhile, these centres are able to draw on larger **trade areas** and greater populations, and achieve threshold levels needed to support higher-order services. The development of supermarkets in market centres has strengthened the centrality of these settlements, but at the same time hastened the closure of village stores.

Larger central places have also been strengthened by rural regeneration policies such as key settlements, which have concentrated schools and healthcare services in market towns. New housing developments have been diverted to key settlements which already have a full range of retail, educational, healthcare and transport services. The service status of villages has diminished in rural commuter zones. Commuters are highly mobile and because they have strong ties to large urban areas through employment, they rely less on village services. In remote rural areas, depopulation **undermines thresholds**, and without government subsidies post offices, general stores and primary schools close. Holiday homes and second homes have also contributed to the demise of services in village settlements.

The decline of small, central places in rural settlement hierarchies has hit those groups who are least mobile, for example old people, the poor who rely on public transport and mothers at home with young children. Higher-income groups who are mobile have been least affected. The lack of services contributes to out-migration, depopulation and a further decline in rural services.

What are the causes of rural service decline? How successfully has rural service decline been tackled?

Rural services are mainly shops, schools, medical/paramedical services and public transport. Service decline was a feature of most rural settlements in MEDCs in the second half of the twentieth century, especially in **lower-order settlements**. The causes of decline include:
- rural depopulation
- competition from urban centres
- **economies of scale** in service provision which make small shops, single GP practices and small primary schools cost-ineffective
- improved personal mobility and faster roads, extending the **ranges, thresholds and trade areas** of higher order settlements
- counterurbanisation, with commuters retaining strong functional ties with urban centres

Key settlement policies have concentrated services in market towns which have the **population potential** to support them. New housing is also located in key settlements which are able to provide the necessary service provision. This policy, adopted by most rural counties in England, has helped to retain a minimum service provision for many rural regions. Government subsidies for rural shops (e.g. rate relief, financial help, and the Sainsbury's SAVE scheme) and protection of schools against closure are crucial to the survival of services in smaller rural communities. Subsidies, community buses and partnership schemes help to provide basic transport links between villages and market towns. Even so, most rural communities are without a village shop, post office or public transport.

To what extent are human activities in national parks sustainable? How successful have governments been in achieving sustainable management?

Human activities in national parks include tourism and recreation, mining and quarrying, water supply, agriculture and military training. The **sustainable use of resources** does not create negative environmental impacts.

Many national parks are victims of their own popularity. Millions of visitors cause overcrowding, traffic congestion and pollution from vehicles. Visitor numbers may exceed the **carrying capacity** of the environment and cause footpath erosion, while visitors straying from official trails may destroy vegetation by trampling on it. Water-based recreations such as jet- and water-skiing also cause noise pollution and are incompatible with quiet activities like fishing. In US national parks, devoted almost exclusively to conservation, sustainability issues centre mainly on tourism and recreation. National parks in England and Wales are not exclusively for recreation — many economic activities, such as farming and quarrying, pre-date the establishment of the parks. The **intensification of farming** in the past 30 years and continued quarrying of rocks, such as limestone, granite and slate, are unsustainable.

Management of national parks aims to achieve sustainable use of resources. Visitor pressure may be countered by promoting honeypot locations. In US national parks, where restricted access allows control of visitors, park-and-ride, using shuttle buses, can solve problems of traffic congestion (e.g. Zion and Yosemite). Footpath repair using durable surfaces helps to arrest erosion, and visitor centres are able to educate the public about the **fragility of ecosystems** and the **environmental consequences** (i.e. destruction of vegetation and erosion) of straying from official footpaths. In national parks in England and Wales, financial incentives through the environmentally sensitive areas scheme promote less-intensive farming, and encourage farmers to use traditional methods of conserving **landscapes**, **habitats** and **wildlife**.

Hazardous environments

Key questions

Key question	Synoptic connections
How important are physical and human factors in causing natural hazards?	Slope systems and mass movements (2680). Drainage basins and river floods (2680, 2683). Global energy budget and hurricanes (2680). Plate tectonics, earthquakes and volcanoes (2680). Coastal management (2683). Human activities in dryland environments (2683). Hazardous environments (2684).
To what extent has rapid urbanisation in LEDCs increased the risks of natural hazards?	Contemporary urbanisation in LEDCs (2681). The consequences of rapid urbanisation, e.g. squatter settlements (2681). Managing urban environments (2684).
How important is an understanding of physical processes (e.g. plate tectonics, mass movements and atmospheric systems) in assessing the risks posed by natural hazards?	Lithosphere and atmospheric systems; the processes responsible for earthquakes, volcanic eruptions, mass movements and hurricanes (2680). Population distribution and population density (2681). River floods and hydrological processes (2681, 2683). Coastal erosion and depositional processes and coastal management (2683). Human activity in dryland environments (2683). Hazardous environments (2684).
To what extent can disaster planning mitigate the effects of natural hazards?	Lithosphere and atmospheric systems; the speed and magnitude of hazard events (2680). Hydrology, runoff, flooding and flood management (2680, 2683). Population distribution and density (2681). Urban and rural populations in MEDCs and LEDCs (2681). Hazardous environments (2684).
To what extent does the impact of natural hazards depend on levels of economic development?	Levels of economic development and natural hazards (2684). Factors other than levels of development that influence natural hazards: population density and distribution (2681, 2684). Accessibility; speed and magnitude of hazard events, e.g. earthquakes, volcanic eruptions, landslides (2680). Hazardous environments (2684).
How does geographical location (e.g. coastal/inland, upland/lowland, urban/rural) influence the vulnerability of areas to hazard events?	Physical environment (2680). Human environment; urban and rural populations (2681). Coastal environments (2683). Fluvial environments (2683). Hazardous environments (2684). Managing urban environments; managing rural environments (2684).

Key questions answered

(Possible synoptic connections are shown in **bold**.)

How important are physical and human factors in causing natural hazards?

Human activity has little or no influence over causing most natural hazards. Hazards such as hurricanes, volcanic eruptions and earthquakes are the result of **physical processes**. Atmospheric processes such as **evaporation**, **condensation**, **latent heat transfer** and the **Coriolis force** cause hurricanes and tropical storms. Volcanic eruptions result from convection flows in the Earth's upper mantle, and processes such as **subduction**, **rifting** and **sea-floor spreading**. Earthquakes are caused by compressive, tensile and shearing forces in the **lithosphere** along **tectonic plate margins** and **fault lines**.

However, human activities can influence some hazard events. Mass movement hazards often result from the interaction of physical and human processes. Physical factors include **slope materials**, **slope angles**, **pore water pressure** and vegetation cover. Human activities that can contribute to slope failure are **loading** (e.g. by building or tipping on slopes), **devegetation** and changes in the **hydrological balance**. Flooding by rivers is also influenced by human activities, especially land use changes which may accelerate **runoff** (e.g. **urbanisation**, artificial drainage and **deforestation**) and reduce **storage** and **evapotranspiration**.

To what extent has rapid urbanisation in LEDCs increased the risks of natural hazards?

Urbanisation is an increase in the proportion of urban dwellers in a country or region. Urbanisation in LEDCs is the result of **rural–urban migration** and **natural increase** in urban areas.

The effect of urbanisation is to concentrate large numbers of people at **high densities**. It can be argued that the impact of natural hazards is directly related to the population at risk, making urban populations more vulnerable than rural ones. Large numbers of urban dwellers in LEDCs are poor and live in **shanty towns**. Shanty towns often occupy sites unattractive to development because of hazard risks such as flooding and mass movement (e.g. favelas in Rio de Janeiro). Flimsy dwellings are also vulnerable to atmospheric hazards such as hurricanes (e.g. Hurricane Mitch in central America).

An alternative view is that urbanisation has reduced hazard risks. It is easier to plan for disaster management in urban areas. Urban areas are more prosperous than rural areas, have a more comprehensive **infrastructure**, and, as a result, are more accessible. Large **urban agglomerations**, such as those in southeast Brazil, are far better placed to plan for and respond to hazard events than **poor rural regions** such as the northeast.

How important is an understanding of physical processes (e.g. plate tectonics, mass movements and atmospheric systems) in assessing the risks posed by natural hazards?

The risks posed by natural hazards depend on both physical and human geography. An assessment of these risks can be considered with respect to specific hazards:
- volcanic eruptions
- earthquakes
- mass movements
- hurricanes
- floods

The risks from volcanic eruptions depend largely on the type of eruption. Eruptions at **constructive plate boundaries** are **effusive**. **Basalt lava** has low **viscosity**, allowing gas to escape freely and minimising the risks of violent eruptions. At **destructive plate boundaries**, **andesitic lavas** are viscous and eruptions are often violent (e.g. Mount St Helens in 1980). Understanding the processes that cause **inflation** of the ground surface (as the magma chamber fills) can help to forecast eruptions. Volcanoes that produce **pyroclastic flows** (e.g. Vesuvius or Soufrière Hills) are particularly hazardous. An understanding of these eruptions can inform planning for disaster management.

The impact of earthquakes depends partly on their magnitude. Shallow earthquakes are more damaging than deep ones. Earthquake depth depends partly on **types of plate margin**. The risks of earthquakes also depend on the tectonic activity along **fault lines** and on the interval between quakes.

Hurricanes derive their energy from evaporation of warm surface water in tropical oceans. Their energy is soon dissipated over land, placing islands and coastal regions at most risk. The movement of hurricanes and likely landfall can be predicted from an understanding of physical processes operating in the atmosphere (e.g. **pressure patterns**, **upper troposphere winds** and the **Coriolis force**).

Accurate flood forecasting requires an understanding of **drainage basin hydrology**, and flows of water between stores. The speed of flow (**runoff, throughflow** and **groundwater flow**) and the status of stores (soil, vegetation and permeable rocks) will largely determine **lag times** and **peak flows**.

Nonetheless, human factors also have a significant influence on hazard risks. **Population distribution and density**, **urbanisation**, **levels of economic development**, **disaster planning** and disaster prevention all have at least as much effect on hazard risks as an understanding of physical processes.

To what extent can disaster planning mitigate the effects of natural hazards?

Disaster planning helps mitigate the effects of natural hazards by:
- educating the population at risk

- designing buildings to minimise the risk of collapse or fires
- preparing emergency plans for evacuation and relief
- monitoring potential hazards and issuing warnings
- building hard structures to divert potential hazards such as floods and lava flows or to protect a population at risk from tsunamis and typhoons

Education may involve practising drills for safe evacuation in the event of earthquakes and tsunamis, and informing people of the appropriate action to take during and in the aftermath of an earthquake. Evacuation plans may include designating 'safe areas' where people can assemble. In MEDCs such as the USA and Japan, modern high-rise buildings in earthquake-prone regions are often stabilised by counterweights, cross-bracing and shock absorbers. Planning regulations in earthquake zones may also require buildings (houses, offices and factories) to be fireproof. Volcanoes may be monitored for **gravity changes**, **inflation** and **seismic activity**, and hurricanes and other tropical storms may be tracked using **satellite imagery**. **River catchments** can be computer-modelled and monitored in **extreme rainfall events** (e.g. in terms of **rainfall amounts**, **soil moisture conditions** and **evapotranspiration**) to provide warning of potential flooding. Hard engineering structures such as **relief channels for floodwater** and lava flows, **sea walls** and cyclone shelters may be built to protect populations against potential disasters following a hazard event.

Despite planning to mitigate the impact of natural hazards, disasters still occur. Even in MEDCs there is significant loss of life and damage to property from natural hazards such as earthquakes (e.g. Kobe 1995 and Taiwan 2000) and floods (e.g. UK Midlands, April 1998). The impact in LEDCs is much more severe, not just from earthquakes and floods but also from volcanic eruptions, tsunamis, landslides and hurricanes. Contrasts in **levels of development** and the impact of natural hazards in MEDCs and LEDCs is one indication of the effectiveness of disaster planning.

To what extent does the impact of natural hazards depend on levels of economic development?

The impact of natural hazards depends on several factors, including:
- economic development
- population distribution and density
- the severity, magnitude or nature of the hazard event
- timing (time of day or year)
- accessibility of the area hit by the hazard

Levels of economic development have a huge influence on the impact of hazards. MEDCs have the wealth to mitigate the effects of natural hazards. Buildings can be made earthquake-resistant and fireproof; **hard structures** such as dykes, barrages and sea walls can prevent coastal flooding; and flood relief channels and dams can reduce the hazards of river flooding. MEDCs have the resources to monitor potential hazards such as volcanoes and hurricanes, and efficient communication systems can provide early warning of impending disasters. MEDCs have **literate populations** and can educate them for disaster contingencies. MEDCs have the resources to access regions hit by natural hazards and provide funding for emergency aid and long-term relief. Finally, **poverty** may force people in LEDCs to inhabit areas that are more vulnerable to hazards (e.g. steep slopes, flanks of active volcanoes and flood plains of rivers).

But the impact of natural hazards is affected by other factors. **High-density urban populations**, whether in MEDCs or LEDCs, are more at risk than **low-density rural populations**. The Kobe earthquake in 1995, which killed over 5000 people, showed that even the richest countries are not immune from major earthquake disasters. The nature of the hazard also influences its impact. A sudden explosive eruption and pyroclastic flow will devastate any community in its path, whether in an MEDC or LEDC. A major earthquake, which is likely to occur without warning, could be equally damaging. Timing affects hazard impact. Earthquakes occurring at night (when most people are indoors) are likely to cause greater casualties than those taking place during the day. If a major hazard strikes during the **winter months**, its secondary effects (lack of food and shelter) will have more impact than in the summer. Access is also important. **Remote areas** are found in MEDCs (e.g. Alaska, USA) as well as in LEDCs, and this may delay emergency aid and relief.

How does geographical location (e.g. coastal/inland, upland/lowland, urban/rural) influence the vulnerability of areas to natural hazards?

Coastal areas are vulnerable to flooding from hurricanes, tsunamis and **storm surges**. Mass movements (rock falls, landslides and mudslides) are common along some **upland coastlines**, but with few people living close to the shoreline, risks are not great. A large part of the **world's population**, and many of the **world's largest cities**, are located on or near the coast. This greatly increases the risks from hurricanes, tsunamis and storm surges. Inland areas are less at risk from hurricanes but have their own hazards. Tornados most often develop in continental interiors and cause severe but localised destruction. Inland areas are likely to have steeper slopes and therefore are exposed to risks of mass movement.

Upland areas are often **high-energy environments**. **Gravity** and vigorous **erosion** may trigger landslides and other mass movements. Volcanoes form some of the highest mountains in the world and active volcanoes pose hazards from lava flows, pyroclastic flows and super-heated gases. **Tectonic activity** is often present in upland areas, increasing the risk of earthquakes. Because many upland areas are remote and **isolated**, emergency aid may be slow to materialise following a disaster, and this may increase the death toll. In lowland regions, river flooding is more likely: rivers have higher **discharge** and large populations inhabiting **flood plains** are particularly at risk.

Urban areas are susceptible to hazards such as earthquakes, hurricanes, volcanic eruptions and mass movements because of their **high population densities**. However, urban areas are more prosperous than rural areas, and disaster planning and communications are more advanced and more sophisticated than in the countryside. In rural areas in LEDCs, few buildings will be earthquake-proof. **Poverty, isolation and remoteness** may all increase the primary and secondary impact of hazard events in these areas.

Questions & Answers

This section contains examples of student answers to eight exam-style questions based on the People and Environment Options unit. There are two questions and answers for each option. All of the questions have a similar style: they are discursive and evaluative; they provide opportunities for synoptic writing; and they have a simple form, unsupported by stimulus materials such as diagrams, sketch maps, charts and photos.

Examiner's comments

Examiner's comments, preceded by the icon *e*, punctuate the sample essays. These point out strengths and weaknesses, and suggest areas for improvement and development. The assessment criteria (see pp. 11–12) guide these critical comments; in particular, there is emphasis on the quality of knowledge and understanding, the use of place and environmentally specific examples, synoptic connections and evaluation. Summative comments are found at the end of each essay, together with the marks awarded for the four assessment criteria. A final grade is awarded to each essay. This is determined by the notional 'designer threshold' boundaries, i.e. 80% and above = A, 70–79% = B, 60–69% = C and so on.

Geographical aspects of the European Union (EU) (I)

To what extent have government farm policies had an adverse effect on the physical environment? Illustrate your answer with reference to the EU.

e A successful answer will explain and exemplify the connection between farm policies and physical environmental change. Appropriate content might include the adverse environmental effects of farm policies promoting increases in output, either by intensification of production or by farming extensively. The former includes the increased use of machinery and agrochemicals such as chemical fertilisers and pesticides; the latter involves the reclamation of land, the conversion of pasture to arable, and the ploughing of downland and heathland. Balanced answers will also cover the environmental dimensions of some farm policies, such as subsidies to encourage less-intensive production and organic farming, and promoting environmental conservation. Detailed examples of specific places and environments in the EU should be a feature of good answers. An effective synoptic approach will draw on material from human, physical and environmental geography.

■ ■ ■

Candidate's answer to question 1

The Common Agricultural Policy (CAP) is the EU's farm policy and was set up in 1962. Agriculture employs 4.7% of the EU's workforce and contributes just 2% of the EU's gross domestic product (GDP). However, it accounts for nearly half of the EU's budget. The main aim of the CAP is to provide food security. It also aims to improve agricultural productivity, produce food at a fair price and prevent excessive fluctuations in the price of food.

e The introduction merely outlines the EU's Common Agricultural Policy. A more effective introduction might (a) define the term 'physical environment' and (b) give an idea of the structure of the discussion to follow. For example, it might make the point that farm policies have both positive and negative effects on the physical environment, and list some of these.

During the 1970s and 1980s, agricultural output from the EU rose rapidly. This was due to the intensification of farming. Inputs to the land, such as chemical fertilisers, pesticides and irrigation water, were increased to give greater yields. Many small farms were taken over by larger ones. These larger farms were able to achieve economies of scale in using machinery and buying inputs such as seed and fertiliser.

Intensive farming was bad for the environment. The more food farmers produced, the more money they received from the CAP. Intensive farming involved the use of chemical fertilisers, especially nitrates. Nitrate is very soluble in rainwater and is easily

question 1

washed out of the soil. Once leached from the soil it gets into lakes and rivers via runoff and groundwater flow. Nitrates in lakes, rivers and groundwater are harmful to the environment as they encourage algal blooms which consume oxygen in the water and destroy aquatic life. As a result, nitrates have ruined many aquatic ecosystems. Also, nitrates in drinking water lead to illness in children.

> *e* These two paragraphs show knowledge and understanding of physical and economic processes and the impact of human activities on the physical environment. The candidate makes connections with aspects of physical geography (e.g. ecosystems and soil) and economic geography (e.g. agriculture). Implicit in these paragraphs is the idea that the interaction of physical and economic systems creates issues of environmental geography.

In Brittany, northwest France, intensive agriculture has also had a harmful effect on the environment. Brittany has a rugged coastline with sandy beaches and is popular with tourists. But the region is also important for agriculture. It produces 14% of France's agricultural output by value. It contains many pig farms. The EU gave grants to pig farmers to develop intensive farming. The huge number of pigs in Brittany has led to pig slurry being washed into the coastal environment through rivers and groundwater flow. Nitrates from chemical fertilisers used to grow cereals for pigs, and ammonia from pig slurry, have contaminated coastal fisheries and caused unpleasant algal blooms.

It has become clear that this intensification of farming is bad for the environment and so some farmers have made their operations less intensive, and have converted to self-sustaining organic farming. This type of farming is better for the environment as it does not use chemical fertilisers and pesticides. In the past, intensive farming has caused the contamination of drinking water, forcing people to rely on bottled water. The newly organic farms grow cereals which are fed to the pigs. The manure produced by the pigs is spread on the fields as natural fertiliser. This type of farming, in contrast to intensive high-tech farming, is much more environmentally friendly.

> *e* Detailed reference to a specific place — in this case Brittany — provides useful exemplification of general physical and economic processes and their environmental impact. The understanding of processes, combined with a specific regional example, provides content drawn from 2684 and from other aspects of geography in the specification.

Although the CAP still subsidises farmers in the EU, it no longer promotes intensive operations. Thanks to set-aside, farmers are even paid to leave their fields empty.

In the UK over the past 25 years, there has been a massive decline in the agricultural workforce. This is partly due to an ageing farm population. When a farmer retires, there is often no one to continue running the farm, so the business ends; small farms are sold off to larger ones. Also, farm incomes, especially on small farms and in the uplands, are often too small. Even with subsidies, many upland farmers in the UK earn as little as £5000 a year. The result is that many small farmers leave their fields empty.

> *e* The final paragraph fails to provide any conclusion or synthesis of the discussion. Indeed, the answer is, in general, rather poorly structured. Its main weaknesses are its lack of

any evaluation, its limited attempts to link discussion explicitly to the question, and insufficient understanding of the connections with different aspects of geography. On the positive side the answer contains some good detail of processes and specific examples, and the content is relevant to the question.

🖉 **The answer is awarded 4/8 for knowledge, 10/22 for understanding, 11/22 for application of knowledge and understanding (including evaluation) and 5/8 for communication. This gives it 30/60, which is equivalent to a D-grade.**

Geographical aspects of the European Union (EU) (II)

To what extent do less prosperous regions depend excessively on localised natural resources? Illustrate your answer with reference to examples from the EU.

> While less prosperous regions may have a number of features in common, there is no single factor which explains their economic problems. Lack of economic success may be due to structural, historical or location problems, as well as an over-dependence on localised resources for farming, industry and services. Good answers will explore in detail the synoptic links between less prosperous regions and their physical and economic environments. This could involve an understanding of how climate, relief and soil influence agriculture in rural regions; how manufacturing industry is affected by natural resources such as energy and materials, as well as by markets, labour and external economies; and how many service activities depend on accessibility, external linkages, skilled labour and cumulative growth. A range of EU regional examples will provide illustration, and discussion should conclude with an evaluation of the question posed.

■ ■ ■

Candidate's answer to question 2

There are five main factors that explain why some regions may lag behind in their development. These factors are indirectly related to a region having an over-dependence on its localised resources. They are historical issues, natural resources, employment structures, cumulative causation and environmental effects.

> This is a brief but effective opening. The candidate flags up the main factors likely to influence regional prosperity. This gives the answer a useful framework.

Despite all the EU's economic advantages, there are many regions lagging behind in terms of wealth and development. The EU has identified two main types of problem region and supports them through its structural funds. Those regions where the GDP per capita is less than 25% of the EU average have Objective 1 status, while those facing structural difficulties, for example in employment, are usually granted Objective 2 status. South Yorkshire is a prime example of a problem region. Its problems stem largely from a previous over-dependence on localised natural resources.

South Yorkshire benefited from the Industrial Revolution because it had a vast amount of coal, which was obviously a vital energy resource, as well as iron ore. Coal was expensive to transport in the nineteenth century, and for this reason energy-intensive industries such as iron and steel were located on coalfields. Iron ore is also bulky and expensive to transport and this also favoured the growth of the iron and steel industry close to its materials (i.e. coking coal, iron ore and limestone).

During the nineteenth century, South Yorkshire became a specialised industrial region and over-reliance on manufacturing was pronounced. With the globalisation of industry, deindustrialisation struck in the 1970s and the region found it hard to adjust. Around 174 000 jobs were lost in coal mining and steel alone. Because South Yorkshire was over-dependent on local natural resources such as coal, there was nothing for it to 'fall back' on. The workforce had few transferable skills and there was little interest in the region from the service sector. Today it is one of the poorest regions in the EU. As a result, it gets assistance from the EU, which will amount to £1.5 billion between 2000 and 2005. The idea behind this assistance is to retrain the region's workforce and stimulate economic growth. The South Yorkshire example shows how over-dependence on natural resources can create regional problems.

e This shows good understanding of the initial advantages of nineteenth-century industry and the role of transport in the location of heavy processing industries. This links to other areas of the specification, such as the manufacturing industry option. Similarly, the processes of globalisation and deindustrialisation provide further connections to economic geography outside the 2684 options.

Other problem regions often rely disproportionately on agriculture. Greece has five of the EU's ten poorest regions. Moreover, Greece is the only country in the EU where more than 10% of GDP comes from agriculture. But Greece's reliance on agriculture and the resources supporting agriculture is in part due to its peripheral location. Being on the periphery, Greece is a long way from the main markets (London, Paris, Randstad and Rhine-Ruhr). This core area is the centre of financial services and has a market of 180 million people. This is where the process of cumulative causation 'kicks in', as service activities are attracted to the biggest markets, which offer external economies, and well developed infrastructures. One result is that regions on the periphery (e.g. Greece, Portugal and southern Italy) lag behind in their development. Often they have few options but to rely on economic activities like agriculture, which use local resources.

e The importance of relative location and access to markets is made, but could be given more emphasis. These factors show that we cannot explain the lack of economic success of less prosperous regions only in terms of an excessive dependence on localised natural resources.

As I have shown, there are complex reasons why regions become dependent on localised resources. A location that is a long way from the core is one such reason. However, not all regions whose development lags behind are over-dependent on natural resources. Norrbotten in Sweden has plentiful natural resources such as hydro-electric power (HEP), iron ore and timber, but it is a remote geographical location which has caused problems for development.

e This is an important example which furthers the candidate's argument. It shows that a region that relies on localised natural resources may owe its lack of prosperity to another factor.

question

Norrbotten is one of the most northerly regions in the EU. It is 1000 km from the Swedish capital, Stockholm, and was the last area in Europe to become free from glaciation. With a population density of just two people per km² (the lowest in Europe), local markets are small and the generation of wealth is a problem.

To sum up, while it is true that many regions that are termed 'problem' do rely on their natural resources, there are many examples where this relationship is untrue. Many things can go against regions. Location is a major factor. So, too, are structural difficulties, depopulation and an inability to attract investment.

✎ Overall this is a good answer, though there is still some imbalance in content between the 2684 option and other aspects of the specification. More knowledge and understanding from the latter would give a better balance. The answer is well structured. It has a clear introduction, provides evaluation in the final paragraph, and offers a considered conclusion. The level of factual detail is good, with place-specific examples and factual information including, for example, the number of jobs lost through deindustrialisation in South Yorkshire and the population density of Norrbotten. There is effective and accurate use of geographical terminology throughout.

✎ **The answer is awarded 5/8 for knowledge, 16/22 for understanding, 17/22 for application of knowledge and understanding (including evaluation) and 6/8 for communication. This gives it 44/60, which is equivalent to a B-grade.**

Managing urban environments (I)

Discuss the view that the continued growth of large cities in LEDCs is unsustainable.

🖉 Sustainable urban development describes a type of growth that could be maintained indefinitely without destroying the resources on which it depends. This question focuses on the relationship between human activities and the physical environment. The rapid growth of cities in LEDCs has had a negative impact on the physical environment, including pollution of water and the atmosphere, and has increased the risks of environmental hazards such as flooding and slope failure. Evidence of unsustainable growth is also provided by inadequate provision of housing, services and employment in many LEDC cities. However, the question offers scope for discussion. There are positive signs of progress through effective management and planning in some LEDCs (e.g. in Latin America) and it would be simplistic to imagine that all LEDCs face the same problems. For example, urban problems in south Asia and Africa are more acute than those in South America or in east and southeast Asia.

■ ■ ■

Candidate's answer to question 3

This essay will identify the effect of the continued growth of large cities and evaluate the degree to which this growth is sustainable. An LEDC is a less economically developed country such as Ethiopia or China. Sustainability refers to the ability to achieve something in the present without foregoing something in the future. Sustainable growth would therefore be growth in cities today that does not prevent growth in the future.

🖉 At the outset, it is essential to define the key term(s) in the question. The candidate provides a sound definition of 'sustainability'.

Over the past 50 years, urbanisation has occurred in LEDCs. The urban population has grown relative to the rural population, due to a combination of natural increase and net in-migration. Of the 28 mega-cities in the world, 22 are now in LEDCs. Mexico City and São Paulo are examples. Both have grown as a result of rural–urban migration and step-migration.

🖉 Effective synoptic responses make connections with other aspects of geography in the specification. In this paragraph, the candidate uses appropriate synoptic knowledge and understanding of demographic processes, such as migration, natural increase and urbanisation, studied at AS (2681).

The growth of these mega-cities has created their own microclimates. In August 2001, the temperature in central London was 3°C higher than in the Surrey countryside south of London. This is the heat island effect.

question 3

In cities in LEDCs, pollution is higher than in rural areas, and higher than in MEDC cities. A student investigation of the River Crane in London showed that the water had a dissolved oxygen content of at least 40% (class II). However, in a shantytown (Grogol) in Jakarta, Indonesia, river water had far less oxygen and a higher lead content. The lead was the result of unfiltered particles from car exhausts, which in the UK are retained in motor vehicles by catalytic converters.

The growth of cities puts huge pressure on any city's ecological footprint. This is the area around a city which supplies it with the resources (physical and human) that it needs. For example, deforestation has occurred in the rainforest surrounding Brasilia (the capital of Brazil), which is unsustainable. Without forest trees to hold the soil in place and intercept rainfall, leaching of nutrients and extensive soil erosion has taken place. This makes the land uncultivable without the use of chemical fertilisers.

> *e* In the preceding three paragraphs, synoptic links with physical and environmental geography (e.g. heat islands, biological oxygen demand, biogeography and ecosystems) can be credited. References to cities in MEDCs (irrelevant in this context) are ignored.

Within cities, the biotic index of plant and animal diversity is much less than in nearby rural areas. In the Kruger National Park in South Africa, over 105 different species of grass have been identified, while in urban parks in Pretoria, only four species are found. This is another example of the pressure of urban growth on the physical environment in LEDCs and how this growth is unsustainable.

Settlement structures in LEDCs have meant step-migration occurring as migrants respond to real and perceived attractions of cities. In addition to these pull factors, push factors such as failed harvests and civil war make people leave the countryside and head for higher-paid jobs in conurbations such as Mexico City, which now has a population of over 30 million. The majority of people who enter the city live in favelas on the outskirts. Housing shortages in Rio de Janeiro have caused people to set up shacks on steep hill slopes, leading to soil erosion and mudslide hazards.

In Johannesburg, South Africa, the brownfield sites left from gold mining contain large waste deposits of lead which contaminate streams in Soweto, causing a decline in water quality. The same happens in MEDCs, such as on the sites of disused gasworks in Greenwich, London. However, in LEDCs there is often no legislation to prevent pollution and so the leaching of heavy metals such as lead continues.

> *e* Further synoptic connections are made, showing sound knowledge and understanding of A-level geography, through references to migration, natural hazards and environmental issues. Again, detail of MEDCs are irrelevant and therefore not credited.

Green belts, set up to prevent urban sprawl around cities, are also at risk. In Lagos, Nigeria, the national football stadium is being rebuilt to support a bid for hosting the World Cup. There is little space and so the stadium is being built on a greenfield site. Like the urban–rural fringe shopping centres of MEDCs (Bluewater, Kent), this causes increased air and noise pollution in the surrounding area.

Tourism has also led to increased pressure in LEDCs. Eco-tourism and bottom-up projects in countries such as Belize and Tanzania are designed to protect the environment and help the poorer people. This, however, cannot always be the case. In Bali, a concrete jungle has developed, and this has led to shortages of water and problems of waste disposal. In Jamaica, a new airport is planned, but a cost/benefit analysis showed that the costs from pollution (air and noise) would outweigh the benefits from visitors, even taking the multiplier effect into account.

e This paragraph makes synoptic connections with tourism (2683).

The development of a country does play a role in the sustainability of growing cities in LEDCs. Corrupt governments and lack of proper legislation prevent the reduction of waste and pollution at a cost to the environment.

There is a strong relationship between the physical environment, which includes the atmosphere, hydrology and ecosystems, and human geography, which includes settlement, population growth and tourism. Together, physical and human factors contribute to the unsustainable growth of large cities in MEDCs.

e This is a wide-ranging answer. The candidate has gone to some effort to make clear the connections between urban development and other aspects of the specification. A range of links between urban development and physical, human and environmental geography are explored. Other qualities in this answer include its factual detail and its discursive and evaluative approach. Positive marking means that irrelevant details are ignored.

e **The answer fulfils the requirements for synopticity. Specifically, it is awarded 7/8 for knowledge, 19/22 for understanding, 20/22 for application of knowledge and understanding (including evaluation) and 7/8 for communication. This gives it 53/60, which is equivalent to a good A-grade.**

Managing urban environments (II)

Using examples, examine the extent to which urban problems in LEDCs have their roots in the rural areas of these countries.

e The urban problems that affect LEDCs include poor and inadequate housing, poverty, lack of services and employment, traffic congestion and environmental pollution. Some of these problems have strong synoptic links with aspects of physical, human and environmental geography. For example, the physical, environmental, economic and social conditions in many rural areas in LEDCs which stimulate out-migration connect to studies at AS in module 2681. Rural environmental problems such as drought and land degradation, and urban environmental problems such as flooding and landslides, provide opportunities to demonstrate synoptic knowledge and an understanding of physical processes. Similar opportunities to explore physical processes include atmospheric pollution, water pollution and disease in urban areas. A balanced view of the question might conclude that problems of rapid urban growth and inadequate resources stem as much from natural increase in LEDC cities as from in-migration and limited investment.

■ ■ ■

Candidate's answer to question 4

We have all heard stories about the cramped, dirty squatter settlements and rapid growth of cities in the developing world. In Cairo alone, 2 million people live in the City of the Dead, an area of ornate mausoleums which have been occupied by squatters. The population of Bogota in Colombia has doubled to 10 million in the past 15 years. Huge favelas on the outskirts of Rio de Janeiro in Brazil, with no sanitation, clean water or electricity, are the perfect breeding ground for diseases like cholera and typhoid. But to what extent are the real problems not in the urban settlements themselves, but in rural areas?

e This introduction states some of the problems affecting cities in LEDCs. It offers a framework for subsequent discussion, which is useful to both the candidate and the examiner.

The rural–urban migration taking place in the developing world is largely due to poor conditions in rural areas. There are many so-called push factors which cause people to migrate to the cities. A good example of where rural–urban migration is happening is the Ceara province of northeast Brazil. The problems faced by the people who live in villages in this region include lack of education and basic medical care — the latter having led to the spread of tropical diseases. In the villages, there is no access to sanitation or clean water and the workers receive a pittance for working on the plantations owned by large landowners and multinational companies (a hangover from colonial times). Few have the skills to escape poverty, and being landless means that they are forced to buy expensive food. Many people are malnourished and cannot afford the high cost of meat, so levels of health are generally poor. Even those who do own a little land are constantly beset with problems of climate. Due to prolonged droughts, the crops fail frequently, making life extremely difficult. Life expectancy is only around 45 years and, typically, infant mortality rates of 200 per 1000 can rise to 500 during times of drought. The promise of jobs in factories in the cities, as well as healthcare and education, drive many to leave and head for the coastal cities of Fortaleza and Recife.

In Colombia, there has been civil war for the last 20–30 years, which has forced people to flee the countryside for Bogota in an attempt to escape the crossfire between rebels and paramilitary groups.

e The candidate shows good synoptic knowledge of the social, economic and environmental problems of rural areas in LEDCs. The place-specific nature of much of this material — e.g. Ceara in northeast Brazil — gives plausibility and depth to the answer.

Many would argue that the cramped conditions in large cities (up to ten people per room, no sanitation or proper housing and high levels of crime) result from the massive influx of people from rural areas. They argue that the real problem lies in the countryside and that the only way to stem the flow of immigrants into shanty towns is to solve the problems of the countryside.

e This is a useful evaluative paragraph which steers the answer back to the question. It is preferable for evaluation to be interspersed throughout an answer and not left to a final paragraph.

However, it is arguable that this is not the whole problem. Successive governments in Brazil, for example, have done little to aid the people living in the shanty towns. While governments have given money to transnational corporations, such as Fiat, to open prestigious new factories in the cities, they have failed to address the massive shortage of jobs and unhygienic conditions in the shanty towns. Large capital-intensive factories do little to help the poor. There is certainly more that governments could do to fight the terrible poverty in cities.

e The discussion considers alternative arguments. References to transnational companies and capital-intensive industries suggest further synoptic links; unfortunately, these have not been developed in detail.

One approach is through sites-and-services schemes. These are relatively cheap and often funded by international organisations and non-government organisations. In these schemes, areas are set aside for homeless people and supplied with piped water, electricity, paved roads, sanitation etc. People are allowed to build their own homes. Sites-and-services have helped improve living conditions in many cities in South America.

A second approach is 'upgrading'. Governments invest in basic infrastructure in existing shantytowns. This approach has been used successfully in cities such as Rio de Janeiro and Cairo. In the City of the Dead in Cairo, a water supply has been added so that people no longer have to walk up to 2 kilometres to collect water.

question 4

Sewage was also a problem, as the original nineteenth-century sewerage system could not cope with waste from 17 million people. The Greater Cairo Waste Water Project helped to alleviate this problem when it was introduced in 1993.

e It is not clear how this discussion of solutions to housing problems relates to the question. For a paragraph or two, the answer loses its way. However, it soon gets back on track.

Governments in LEDCs have often taken inappropriate measures to house in-migrants. In Cairo, large new towns, such as Tenth of Ramadan City, have been built outside the main centre, but have not been particularly successful because people cannot afford to rent the flats there. Large areas of apartment blocks have been built on the edge of the city without places of work. People are too poor to commute from such peripheral areas.

Lack of capital also hinders efforts to improve conditions in the slums. Governments are often in heavy debt to banks and MEDCs and so do not have the resources to develop even the most basic education and primary healthcare for their citizens.

It is a vast oversimplification to say that all the problems found in urban areas in LEDCs are attributable to the rural areas of these countries. In conclusion, it would be fair to say that were it not for the massive rural–urban migration in LEDCs, the problems of overcrowding and poverty would not be as bad. But other factors have contributed to urban problems, such as governments' inability or unwillingness to act to help the poorest. However, the terrible living conditions in the countryside are probably the main driving force behind urban problems.

e The answer has some weaknesses. For example, the range of problems considered could be wider, and in one or two places the relevance of the discussion is not clear. But, overall, this is a good answer: it is discursive, evaluative and makes some appropriate synoptic connections. The answer concludes with a very effective synthesis of the preceding discussion.

e **The final assessment is 51/60 — equivalent to an A-grade — which comprises 6/8 for knowledge, 19/22 for understanding, 19/22 for critical application of knowledge and understanding and 7/8 for communication.**

Managing rural environments (I)

'Rural population changes in MEDCs in the past 30 years have generated important economic and social issues.' With reference to named rural areas, describe these major issues and discuss the effectiveness of planning responses to them.

e Rural population changes include counterurbanisation in pressured rural areas and depopulation in remoter rural areas. One effect of counterurbanisation is 'population turnover' which describes changes in the demographic, social and economic structure in rural populations. A clear description and explanation of these population changes will demonstrate synoptic understanding of the connections between population, social and economic geography. The issues generated by rural population change include:
- poor service provision
- a lack of affordable housing
- social exclusion
- unemployment
- inaccessibility and isolation
- loss of community
- loss of social cohesion

Planning responses are extensive. Among them are:
- key settlement policies
- subsidies for rural services such as village shops and bus services
- diversification of farming into 'off-land' enterprises supported by the EU's Structural Funds
- the creation of Rural Priority Areas
- various partnership schemes

Good answers will show a sound understanding of some of these planning responses and will evaluate their effectiveness.

■ ■ ■

Candidate's answer to question 5

Perhaps the most well-known population trend over the past 30 years with regard to rural areas is that of counterurbanisation. This trend extended even to the USA for the same reasons of increased mobility, a change in working lifestyle and a desire to live in more tranquil areas. However, as with some parts of the UK, this trend did not extend to all rural areas, such as rural communities in the Appalachians and mid-west farming communities.

e The answer begins with counterurbanisation, which is a process relevant to this question. However, the introduction would provide a more meaningful structure if it also referred to depopulation and population turnover, and listed some of the more significant economic and social issues that affect rural areas.

question

Initially, counterurbanisation was most evident in rural areas that were close to urban areas. One example of this is the village of Buriton, which is 25 miles from Guildford, 12 miles from Petersfield and a 1-hour train journey from London. Buriton has now become a commuter village and this has had profound demographic, social and economic impacts. Demographically, the traditionally close-knit rural population has been replaced by an upwardly mobile professional population. Due to the increased mobility of the population, the village shop was forced to close in December 1999 and the post office soon followed. In addition, the village school has closed and there was a decline in transport services, as is the case in many other rural settlements. The Countryside Commission did in fact cite that in 1999 service decline in rural settlements was even worse than this. The problem for the people who have remained in Buriton, especially the elderly, is that of possible social exclusion which can be defined as not having the ability to benefit from those things needed for a reasonable quality of life.

> The answer shows detailed and accurate understanding of counterurbanisation and its impact. This is an important synoptic link, with the candidate drawing on knowledge and understanding studied through rural population and rural settlement at AS.

While much of the housing in settlements such as Buriton has been improved or gentrified, there is little doubt that house prices have increased dramatically. This has led to the now widespread problem of affordable housing across many rural areas in the UK and other MEDCs. This has not just been due to 'population turnover' (i.e. the influx of a new type of population) but has also been due to the rise in the number of second homes. This is seen throughout MEDCs. For example, the village of Grimaud near St Tropez has 40% of its houses as second homes. The extent of second homes and the problems they can cause is shown in the Lake District. In northwest England, there are 32 000 second homes, many of which are in the Lake District. There is thus a problem of affordable housing for local people and, as well as demographic change, the economy of the area has also been damaged. While a few employment opportunities have appeared, for example for builders and architects, jobs are becoming increasingly seasonal, while former close-knit communities, such as Ambleside and Grasmere, are being torn apart.

> The issue of second/holiday homes is appropriate to the question, though the synoptic links with population change/population turnover could be made more explicit. References to place-specific examples and environments will be credited by the examiner.

However, while counter-urbanisation and second-home ownership have expanded to less accessible and more rural areas, depopulation is still occurring in the remotest rural areas. The village of Allenheads in Northumberland, for example, lost two-thirds of its population following the closure of its lead mine in the 1960s. It, too, suffers from a serious lack of services, as can be seen by the closure of its convenience store in 1985, followed by its inn in 1986. Even a brief dabble with tourism has, to the present day, not prevented a severely declining economy. Thus, one must remember to make clear the distinction between different types of rural area.

🖉 Rural depopulation is mentioned (at last!) and its effects on service provision are briefly stated. Unfortunately, opportunities to explore this synoptic relationship in more detail, showing how declining population levels undermine thresholds for services such as post offices, food shops, general practice surgeries and primary schools, have been ignored.

In response to the issues raised by population changes in many rural areas in MEDCs, we are awaiting the effectiveness of numerous schemes, as many of the problems have only recently been addressed. In the Lake District, many attempts have been made via Section 52 of the Town and Country Planning Act to put owner-occupancy clauses into housing contracts. However, all too often developers ignore them and continue to fail to meet the demand for affordable housing for locals. In the development of Beck Close in Coniston, for example, out of 11 houses built in the 1990s only one was purchased by a local. In a similar development in Coniston, named St Martin's Court, the majority of locals could not afford the high prices. However, within the November 2000 Rural White Paper ('A fair deal for the countryside'), provisions were made for the building of 3000 affordable houses each year in rural areas. In addition to this, further funding was proposed, such as £60 million for the rural bus challenge to help with transport service decline (which can often lead to social exclusion). Funding totalling over £30 million was proposed to help Regional Development Agencies provide rural employment because of the lack of employment caused by the aforementioned population and linked socio-economic changes in rural areas. Lack of employment was also due to the decline of agricultural employment, caused by such factors as the rise of agribusiness and increased mechanisation.

Even corporations are getting involved in helping rural areas. Sainsbury's, for example, has decided to initiate its SAVE scheme (Sainsbury's Assisting Village Enterprises) to help with the provision of village shops which will benefit all rural settlements from Buriton to Allenheads.

It seems that the planning responses to many issues in rural areas in MEDCs have come fairly late in terms of trying to sustain traditional rural populations. In this way, the effectiveness of such schemes as SAVE will only be judged in time. Unlike the Section 52 owner-occupancy clause (part of the previous government's policy), any new planning responses should probably be more strongly implemented to improve their chances of success. The Rural White Paper of 2000 appears to address the major issues that lead to social exclusion and depopulation in many rural areas, though it is questionable whether all this has come too late.

🖉 There are valid comments and descriptions of some planning policies, and the concluding paragraph provides an evaluation. Overall, the answer deals successfully with population change, social and economic issues, and planning responses. There are a number of synoptic links made and the candidate shows knowledge and understanding of these links. However, to achieve the highest grade, the candidate would need to show more in-depth knowledge and understanding of these connections and make them more explicit. The answer is clear and well written, though it lacks an effective introduction. Throughout, the candidate displays detailed knowledge and understanding of specific places, environments and policies. This is a good answer but it falls short of a

question

grade-A mark because the synoptic connections between demographic change and social and economic geography have not been fully developed.

e **The answer is awarded 6/8 for knowledge, 17/22 for understanding, 17/22 for the application of knowledge and understanding and 7/8 for communication. This totals 47/60, which is equivalent to a top B-grade.**

Managing rural environments (II)

'Human activities in national parks in MEDCs often have damaging effects on the physical environment.' Illustrate this statement with reference to national parks you have studied and discuss the effectiveness of management strategies to protect the environment.

> Human activities in national parks always include recreation, tourism and conservation. Depending on the characteristics of national parks, other human activities can include mining and quarrying, forestry, the supply of water, farming and military training. These human activities may be unsustainable and have damaging effects on natural and semi-natural ecosystems. Management responses are varied. Visitor numbers may be controlled, private vehicles banned, and sensitive areas placed out-of-bounds. In heavily pressured areas, erosion control may be undertaken. Occasionally, environmental considerations may not take priority, for example in the development of mineral and energy resources or the provision of employment for local residents.
>
> The question focuses on a key theme of synoptic geography — the relationship between people and the physical environment. Good answers will exploit the many opportunities to show synoptic understanding of the links between human activities in national parks and aspects of physical, human and environmental geography.

■ ■ ■

Candidate's answer to question 6

Rural areas in general face an enormous dilemma between protection of the environment and the generation of income. It is a question of economic versus environmental sustainability. National parks in MEDCs face enormous problems with tourists, who show little respect for the environment but who, on the other hand, generate important income.

> This introduction 'sets the scene' but it neither defines key terms such as 'national parks', 'physical environment' or 'management strategies' nor outlines the structure of the discussion to follow.

A national park I have studied which faces problems of human activities and their damaging effects on the physical environment is the Lake District in Cumbria. The park covers 2292 km² with a population of approximately 42 000. It was designated a national park in 1951 and has its own National Park Authority, which manages the park and makes sure there is a balance between the environment and tourism. With 12 million visitors a year, this is an extremely difficult task. The Lake District itself provides a range of activities attracting visitors, 62% of whom stay overnight. They can go hiking, walking, rock-climbing and bird watching, and also visit local attractions such as the Floating Theatre in Keswick. However, the Lake District National Park has tried to take measures to reduce the negative impact of tourism.

A2 Geography

question 6

e This section provides excellent case study detail but there is little or no attempt to develop synoptic connections.

One of the principal activities for tourists in the Lake District is water sport. The Lake District authorities have aimed to concentrate all water sport activities on Lake Windermere, England's largest lake. This has been a partial success. Noise and pollution are not a problem on other lakes in the national park, as boating is banned. However, there are significant problems with powerboating on Lake Windermere. First, powerboating creates a lot of noise and disturbance in the water, affecting local wildlife, such as breeding birds. Also, powerboats emit pollutants in the water and this has led to eutrophication. Algae, which grow as a result, have contaminated the water. However, the National Park Authority has introduced a zoning scheme. It is not the National Park's policy to confine water sport activities such as sailing, canoeing and wind-surfing to Windermere. Water skiers have to stay in particular zones such as the area around the landing at Bowness. Although many problems like noise have been reduced, pollution still remains a problem. A complete speedboat ban has been proposed but this has caused outrage among locals who get an income from this activity. Again, the conflict between environmentalists (conservationists) and those seeking to make a profit is the story of the area.

e This discussion of the issue of powerboats on Windermere, and the resulting noise and pollution, shows an understanding of links between this option and environmental geography. However, there are inaccuracies (e.g. powerboat pollution does not lead to eutrophication). There is also scope for more in-depth analysis of the links with environmental geography. There is an attempt to evaluate some of the planning policies of the National Park Authority.

In other areas of the national park, hiking, walking and rock climbing have led to the erosion of footpaths, littering and harm to wildlife and domestic animals. The Park Authority has addressed these problems using a zoning management strategy. Quiet areas have been designated where tourists and cars are prohibited. This means there is no threat to the environment in these areas, and here the policies have been a success. However, as a compromise, the area around Lake Windermere in the south and the northeastern area near Penrith have been designated caravan areas. Here there is intense tourism. This zoning policy has been effective and the National Park Authority has limited environmental problems, such as noise, to concentrated areas, relieving the pressure on other areas of the park. Thus, the overall zoning policy in the Lake District National Park has been effective, though inevitably there are still some problems associated with tourism.

e Knowledge and understanding of planning policies are handled competently, though again more detail and accuracy are needed. The discussion is appropriate and evaluative, though synoptic links are not given sufficient prominence.

In conclusion, the Lake District National Park has experienced damaging effects on the physical environment as a result of human activities. However, effective management strategies such as zoning have helped to reduce these damaging effects.

OCR (A) Unit 2684

e The answer is limited in its scope by its focus on a single national park. In this example, extra depth does not compensate for the breadth of problems and strategies which might be achieved by considering two or three parks. Given its somewhat narrow focus, the answer does provide some impressive place-specific detail on environmental problems and management policies. In addition, the candidate has produced an answer that is both discursive and evaluative. The main weakness of the answer is its failure to engage in substantial synoptic discussion. For example, the environmental problems of tourism, such as eutrophication of Windermere, loss of habitat, footpath erosion and traffic congestion, could be amplified to show understanding of the connections between environmental, physical and human geography.

e **The answer scores 5/8 for knowledge, 15/22 for understanding, 18/22 for the application of critical knowledge and understanding and 6/8 for communication. The total score is 44/60, which is equivalent to a solid B-grade.**

A2 Geography

Hazardous environments (I)

'Countries at most risk from disasters caused by natural hazards tend to be poor.'
Discuss, with reference to actual hazard events.

e Natural hazards and ensuing disasters affect all countries, regardless of wealth. Good answers will draw attention to a range of factors (including levels of development) which influence the scale of disasters. Levels of development have a direct effect on disaster planning, including monitoring potential hazards, building regulations, education and evacuation procedures. Other influences, which have little relationship to levels of development, are:
- the magnitude of the hazard
- the distribution and density of population
- the timing of the hazard event and its predictability

The Kobe earthquake of 1995 shows that MEDCs are far from invulnerable to disasters caused by hazard events.

An effective answer will be evaluative and present a balanced discussion. Knowledge and understanding of the connections between natural disasters and aspects of physical, human and environmental geography will be explicit. Examples of synoptic connections include knowledge and understanding of the physical processes underpinning hazards, the importance of population distribution and density, and levels of economic development on a variety of scales.

■ ■ ■

Candidate's answer to question 7

On the one hand, when looking at this statement, I tend to agree that LEDCs are more at risk from disasters and natural hazards than MEDCs. For example, in Ethiopia famine caused by drought was exacerbated by a war which had crippled the country economically. The impact of the Mozambique floods in 1999 and 2000, when the Zambezi River flooded, was made worse because only five helicopters were available to evacuate flood victims.

e It is generally not a good idea to state one's position on an issue at the outset. The introduction should be used to define terms such as 'disaster' and 'natural hazard' and set out the structure of the discussion to follow. In this first paragraph, it is not clear how the examples of Ethiopia and Mozambique link to the question.

In 2001, Mt Nyiragongo near Goma, in the Democratic Republic of Congo (DRC), erupted and displaced 500 000 people due to the rapid flow of basic lava down the sides of the ash cone. Despite the fact that the volcano has erupted nearly every decade in recent history (most noticeably in 1977 when 22 000 people were killed), the 2001 eruption was not predicted. Mt Nyiragongo is located in the East African Rift

Valley and so eruptions are notoriously violent and dangerous. Casualties were made worse because the country's four hospitals were destroyed in the eruption. MEDCs sent in aid to support the DRC in the short term, including £2 million from the UK.

e The answer provides convincing detail on the Mt Nyiragongo eruption. There is accurate use of terminology such as 'ash cone' and 'basic lava' and understanding of the relationship between the nature of eruptions and hazards posed.

Despite the eruption and the disastrous impact it had on the DRC, many other recent volcanic eruptions have been predicted. For example, in 2000 Mt Popacatepetl in Mexico erupted. However, there were no casualties as a hazard map had been drawn up and the area was evacuated 2 days before the eruption, thus saving thousands of lives. Methods of prediction include monitoring the amount of gases emitted, because the composition of chlorine in the atmosphere tends to change immediately before an eruption. Eruptions can also be predicted by seismic monitoring. Tilt meters also measure changes in the angle of the volcano as the ground tends to swell before an eruption, signifying magma rising within the volcano. This shows that eruptions can be predicted. The impact of volcanic hazards is related to a country's economic wealth, because these prediction methods are unaffordable in many LEDCs, including the DRC.

e The candidate advances the argument by showing that monitoring volcanic eruptions is less effective in poorer countries, thus increasing risks. The answer shows good synoptic understanding of physical processes.

El Salvador is another example of a country whose economy has been struggling for many years. This problem was exacerbated by recent civil wars. Short-term development strategies included deforestation in order to export timber and make money. Housing was built on deforested slopes, some of which exceeded the critical 32° angle of repose. The lack of vegetation cover also removed the binding effect of tree roots holding the soil together. Therefore, after an earthquake in 1999, landslides were triggered, killing thousands of people living on the slopes and destroying thousands of flimsy shacks. Damage was especially severe on the Pan American Highway. This example shows how the level of development in a country and the education of its people affect the impact of natural disasters.

However, as shown by the impact of Hurricane Mitch in 1998, the impact of natural hazards is not only related to the wealth of a country. Coastal regions are particularly vulnerable to hurricanes which are formed over the oceans (within 5° of the equator). Hurricane Mitch caused widespread damage throughout the Caribbean, and especially in Kingston, Jamaica's capital.

e There is good use of detailed case studies, covering a range of hazards, in the preceding paragraphs. An attempt is also made to discuss and evaluate the accuracy of the statement in the question. But while slopes and the disequilibrium caused by human activities can be credited as synoptic, more could be done to examine the synoptic relationships between slopes, hurricanes and hazards.

question

A2 Geography

The Kobe earthquake in Japan in 1995 was the most expensive natural disaster in history, despite taking place in a rich MEDC. This was because the most common cooking method was on open gas stoves, which caused fires to spread rapidly after the earthquake. Much of the city of Kobe is built on soft rock and reclaimed land, which amplified earthquake shock waves. The epicentre was also close to the main centre of population, worsening the effects, as well as the focus being shallow — only 14 kilometres below the surface. Overcrowding, and the poor construction of many older buildings, made evacuation difficult. These factors all contributed to the immense impact of the Kobe earthquake, which was largely unrelated to the wealth of Japan.

e This is another relevant example. The final sentence makes a connection with the question.

In conclusion, I think that it is fair to say that, in general terms, the wealth of a country has a large impact on the effects of natural disasters. However, there are other contributing factors to the impact of natural disasters, including geographical location, population density, building structures, level of education and rock types.

e This answer provides a succinct conclusion and underlines the candidate's balanced approach to the question. The answer is impressive in its range of case studies and their detail. Most of the material presented is appropriate and contributes effectively to discussion, though the quality of expression is not as clear as it could be. This is an evaluative and discursive answer, which shows some sound understanding of the connections between hazards and physical geography. However, the synoptic connections with human and environmental geography are relatively weakly developed.

e **This essay is awarded 6/8 for knowledge, 19/22 for understanding, 20/22 for critical application of knowledge and understanding and 6/8 for communication. This gives a score of 51/60, which is equivalent to a sound A-grade.**

Hazardous environments (II)

How important is plate tectonics in any assessment of the risks posed by earthquakes?

e The risks posed by earthquakes depend on many factors. Plate tectonics (e.g. the type of plate boundary, intensity of activity, depth of earthquake foci etc.) is just one factor in the equation. Among other factors are disaster planning, population density, accessibility, levels of development and so on. Good answers will consider plate tectonics and a range of these other factors. They will draw on knowledge and understanding of the connections between earthquake hazards and physical, human and environmental geography studied elsewhere in the specification. Discussion of place-specific examples and actual hazard events should be given prominence.

■ ■ ■

Candidate's answer to question 8

Earthquakes are caused by the build-up of stress in the Benioff zone of lithospheric plates. When this stress is released, a seismic wave is produced. This wave is responsible for the primary hazards of ground movement and shaking, and leads to secondary hazards such as soil liquefaction, landslides and avalanches.

e The answer begins with an explanation of earthquakes. While this is appropriate, some indication of the range of factors influencing earthquake risks should appear in the introduction.

The Mexico City earthquake of 1985 was amplified four to five times by the location of the city on lake sediments. This geophysical factor was largely responsible for the 10 000 deaths and 50 000 injuries, more so than the movement of the lithospheric plates. Thus the earthquake's magnitude was affected by other factors. Another factor that altered risk was the human one. Mexico City is a primate city with a high proportion of its inhabitants in illegal squatter settlements. This is due to rural–urban migration caused by pull factors such as perceived high levels of employment in the city and push factors such as declining agriculture and the fragmentation of land holdings.

e The candidate shows understanding of the connections between earthquake hazards and some aspects of physical and urban geography. This can be credited as part of a synoptic assessment.

The high population density, particularly in squatter settlements (with little effective community leadership or government), was responsible for increased vulnerability, both to the hazard itself and to the ability to recover from the hazard event. Risk assessment, based on vulnerability to a hazard, suggests that Mexico as an LEDC is most at risk from hazards because of human rather than physical factors. The availability of resources such as emergency services following a hazard, and education of

the population before a hazard, are limited as a result of Mexico's relative poverty. An example of this is the cholera epidemic that followed the earthquake. It was caused by a lack of clean water and poor sanitation.

> 🅔 The candidate shows an ability to write discursively and makes the point that in the Mexico example, human factors were more important than physical factors.

Two earthquakes of similar magnitude — Kobe in Japan and Loma Prieta in the USA — had very different effects. This demonstrates some of the factors involved in risk assessment apart from plate tectonics. Kobe in Japan suffered a quake of 7.2 on the Richter scale in 1995. The focus of the quake was quite shallow — 3600 people were killed and over 35 000 injured when the earthquake struck at 5.46 a.m. School earthquake drills in Japan take place four times a year and Kobe's central business district is composed largely of aseismic buildings, with steel frames, deep foundations, crossbeams and sometimes suspension dampers. Most deaths occurred in the Nagata ward where building collapse was most frequent due to timber frame and tile constructions. The Hanshin expressway also collapsed, the elevated sections crushing motorists.

The Loma Prieta earthquake in the USA in 1989 measured 7.1 on the Richter scale. Thirty-six people were killed and 4000 injured when the seismic wave struck in the afternoon. Mapping of the most vulnerable areas had taken place around the San Andreas fault after the 1971 earthquake. Thus, similar earthquake events, which in terms of plate tectonics had similar magnitudes and recurrence intervals, and occurred in countries with similar wealth and education, had very different impacts. This was mainly due to the timing of the two earthquakes.

> 🅔 Although the Kobe and Loma Prieta earthquakes had similar magnitudes, the geophysics of the two quakes are not necessarily comparable. Kobe is adjacent to a subduction zone while Loma Prieta is along the Californian transform fault (San Andreas). Differences in the timing of the two quakes does not account entirely for their contrasting impact. However, the candidate's discursive approach is impressive.

In conclusion, an understanding of plate tectonics is important in the assessment of risk and risk management. For example, areas far removed from plate boundaries, such as the UK, are at low risk from earthquakes compared with those located above Benioff zones. An understanding of plate tectonics may also allow limited prediction of earthquakes. For more predictable hazards, such as volcanoes, evacuation programmes may be put in place and reduce risk significantly (e.g. the evacuation of Montserrat in 1995). However, on plate tectonic boundaries, the interrelationship between geophysical and human factors has a more important role to play in risk assessment — for example, in the case of the Mexico City quake in 1985 or the high population density around Osaka Bay and its significance in the Kobe quake in 1995. The location of these two events led to significantly greater vulnerability for the populations.

> 🅔 The answer ends with an evaluative and summarising conclusion. Throughout, the candidate provides a balanced discussion of plate tectonics and other factors influencing

earthquake risks. Overall, the answer is well constructed, includes synoptic input and its arguments are consistent. However, there is scope for further development (the answer is relatively brief), especially in the area of synopticity.

☒ **The answer scores 6/8 for knowledge, 18/22 for understanding, 18/22 for critical application of knowledge and understanding and 7/8 for communication. The total score of 49/60 is equivalent to a marginal A-grade.**